FOUR
LOST
CITIES

ALSO BY ANNALEE NEWITZ

Nonfiction

Scatter, Adapt, and Remember:
How Humans Will Survive a Mass Extinction

Fiction

Autonomous

The Future of Another Timeline

FOUR
LOST
CITIES

A SECRET HISTORY
OF THE URBAN AGE

Annalee Newitz

W. W. NORTON & COMPANY
Independent Publishers Since 1923

For information about permission to reproduce selections from this book,
write to Permissions, W. W. Norton & Company, Inc., 500 Fifth Avenue,
New York, NY 10110

For information about special discounts for bulk purchases, please contact
W. W. Norton Special Sales at specialsales@wwnorton.com or 800-233-4830

Manufacturing by LSC Communications, Harrisonburg
Book design by Daniel Lagin Design
Production manager: Julia Druskin

Library of Congress Cataloging-in-Publication Data

Names: Newitz, Annalee, 1969– author.
Title: Four lost cities : a secret history of the urban age / Annalee Newitz.
Other titles: 4 lost cities
Description: First edition. | New York, NY : W.W. Norton & Company, [2020] |
    Includes bibliographical references and index.
Identifiers: LCCN 2020012246 | ISBN 9780393652666 (hardcover) |
    ISBN 9780393652673 (epub)
Subjects: LCSH: Extinct cities. | Çatal Mound (Turkey) | Pompeii (Extinct city) |
    Angkor (Extinct city) | Cahokia (Ill.)—Antiquities.
Classification: LCC CC176 .N49 2020 | DDC 930.1—dc23
LC record available at https://lccn.loc.gov/2020012246

W. W. Norton & Company, Inc., 500 Fifth Avenue, New York, N.Y. 10110
www.wwnorton.com

W. W. Norton & Company Ltd., 15 Carlisle Street, London W1D 3BS

1 2 3 4 5 6 7 8 9 0

*This book is given as a humble offering to Iaso,
Acesco, Hygieia, and Panacea.*

*But most importantly it is dedicated
with love to Chris Palmer, who survived.*

CONTENTS

PART FOUR: **Cahokia** THE PLAZA

FOUR
LOST
CITIES

CAHOKIA
1050–1350 CE

POMPEII
700 BCE–79 CE

FOUR
LOST
CITIES

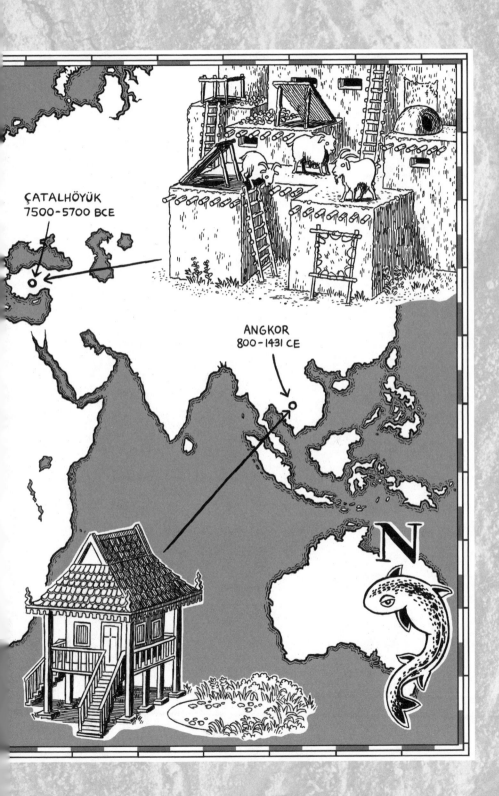

ÇATALHÖYÜK
7500–5700 BCE

ANGKOR
800–1431 CE

N

INTRODUCTION

How Do You Lose a City?

I stood on the crumbling remains of a perfectly square island at the center of an artificial lake created by hydraulic engineers 1,000 years ago. Sunlight played over an eroded sandstone wall. Though this was the dry season in Cambodia, unseasonable rain storms had cleared the air of smoke from local farmers' annual field burning. In the distance, I could see the sculpted towers of Angkor Thom and Angkor Wat, architectural marvels of the ancient capital of the Khmer Empire. Boasting nearly a million residents at its peak, Angkor was once the world's most populous city. And I stood near its center. Beneath my feet was the Mebon, an 11th-century Hindu temple-island, built during the reign of King Suryavarman I in the middle of an enormous reservoir called the West Baray. That morning, the baray's southern shore was dotted with a few motor-boats whose pilots would take visitors out to the Mebon for a couple bucks. It isn't a short journey: the rectangular West Baray is eight kilometers long, roughly the length of three jet runways at a typical airport. A millennium ago, when workers finished digging the baray,

the Mebon temple at its heart was the sole patch of dry land for kilometers around.

Behind its ornate stone gates, the Mebon enclosed another,
smaller reservoir, invisible to all but the select few who were permitted to land on its shores. At its center floated a 6-meter-long bronze
statue of Vishnu reclining, enormous head resting on one of his four
arms. Pilgrims traversed waters within waters to pay homage to this
Hindu god, who brought forth life from the sea when the world was
made. You might say the Mebon is a monument to the spiritual power
of water. But it's also testimony to the ingenuity of Angkorian laborers who kept the annual monsoon floods contained with large reservoirs like the West Baray, and slaked the city's thirst during the
dry season with a system of canals that diverted water from distant
mountain rivers.

Surrounded by shimmering water and weathered blocks from the
temple excavation, I tried to imagine looking out over the baray centuries ago, seeing festive boats full of Khmer locals and dignitaries
from neighboring kingdoms bearing fragrant bundles of flowers and
incense. It must have been astonishing, I thought. But my romantic
fantasy about this place didn't last long.

"I can't believe how much they screwed this up," Damian Evans
said, gesturing in frustration at the baray. Evans is an archaeologist
with the French Institute of Asian Studies whose work over the past
two decades has dramatically changed our understanding of Angkor's
urban grid. A sandy-haired Australian with a quick smile, he's spent
decades writing about the sophistication of the Khmer Empire. But
he's also keenly aware of its failures.

Evans pointed to a faded map of the landscape on a wooden
placard next to us, part of a display detailing a reconstruction of the
Mebon that's currently underway. Looking at elevations, it was obvi-

ous that the east-west oriented rectangle of the West Baray slopes gently with the landscape, causing the east end of the reservoir to fill while the west end stayed dry. As a result, the baray rarely looked like the shining rectangular lake I'd imagined. Instead it would have been more like a deep pool that trailed off into a ragged, muddy edge. But this wasn't because Khmer's engineers were incompetent. "They could have built on a level surface, but the king wanted his engineers to stick with an east-west orientation that suited his gurus," Evans explained. The Khmer believed that grand structures like the king's reservoir should be oriented along the same trajectory that the sun and stars took across the sky. Put another way, King Suryavarman I cared more about auspicious astrological signs than good hydroengineering. This reservoir was an ancient boondoggle. Over the long term, the West Baray became a template for Angkorian city planning, leaving its swelling population with faulty water storage during turbulent periods of climate crisis.

If you substitute the word "politics" for "astrology," Evans' observation could have been made about the design of any number of cities over the past 1,000 years. City leaders pour resources into beautiful spectacles for political reasons, rather than providing good roads, functioning sewers, relatively safe marketplaces, and other basic amenities of urban life. As a result, cities may look awe-inspiring but aren't particularly resilient against disasters like storm floods and drought. And the more a city suffers from the onslaughts of nature, the more contentious its political situation becomes. Then it's even harder to repair shattered dams and homes. This vicious cycle has haunted cities for as long as they've existed. Sometimes the cycle ends with urban revitalization, but often it ends in death.

At Angkor's height in the 10th and 11th centuries, its kings controlled thousands of workers. These were the people who built the cities' palaces, temples, roads, and ill-conceived canals. Though most of this construction was intended to glorify the Khmer kings, it also helped ordinary inhabitants thrive as farmers even during the dry season. But in the early 15th century, the region was stricken by drought, followed by catastrophic[1] flooding that destroyed Angkor's poorly designed water infrastructure at least twice. As the city began to fall apart, the chasm between its rich and poor grew wider. Within decades, the Khmer royal family moved their residence from Angkor to the coastal city of Phnom Penh. This was the beginning of the end for a city whose kings had dominated vast parts of Southeast Asia for centuries, including today's Cambodia, Thailand, Vietnam, and Laos. The city's population had drained away from Angkor's downtown by the 16th century, leaving behind small villages and farms surrounded by Angkor's decaying urban grid. The kings' palaces were abandoned, and the barays became mere depressions in the leafy forest floor. Only a skeleton crew of monks remained to care for the Khmer Empire's legendary temples.

In the 19th century, a French explorer named Henri Mouhot claimed he'd discovered the "lost city" of Angkor. Though other European visitors of the period acknowledged that monks still lived in the Angkor Wat temple enclosure, Mouhot wrote a popular travelogue that suggested he was the first to stumble upon a lost civilization. No human had seen it in centuries, he claimed, and it was full of picturesque wonders to rival those of ancient Egypt. It was an easy myth to perpetuate. Westerners hungry for adventure stories were eager to believe Mouhot when they saw pictures of the city's dramatically collapsing temples, the stones in their walls forced apart by bulging tree roots. From the beginning, Angkor's status as a lost city was manufactured by the media, despite all evidence to the contrary.

The "lost city" is a recurring trope in Western fantasies, suggesting glamorous undiscovered worlds where Aquaman hangs out with giant seahorses. But it's not just a love of escapist stories that makes us want to believe in lost cities. We live in an era when most of the world's population lives in cities,[2] facing seemingly unsolvable problems like climate crisis and poverty. Modern metropolises are by no means destined to live forever, and historical evidence shows that people have chosen to abandon them repeatedly over the past eight thousand years. It's terrifying to realize that most of humanity lives in places that are destined to die. The myth of the lost city obscures the reality of how people destroy their civilizations.

This book is about that reality, which we'll explore in four of the most spectacular examples of urban abandonment in human history. The metropolises in this book all met unique ends, but they shared a common point of failure. Each suffered from prolonged periods of political instability coupled with environmental crisis. Even a powerful, densely populated city like Angkor couldn't survive the double blow of burst dams and chaos in the royal court. Unable build a future in these troubled places, urbanites uprooted their lives and turned their backs on their homes, often at great personal cost. These cities didn't disappear like Atlantis, sliding abruptly below the water into the realm of legend. They didn't go missing. People deliberately abandoned them, for good reasons.

The first city we'll explore in this book, Çatalhöyük, was founded roughly 9,000 years ago during the Neolithic, at the moment when humanity settled down into agricultural life after living as nomads for hundreds of thousands of years. Today its enigmatic remains lie buried beneath two low hills in the Anatolian region of central Turkey. Though small by modern standards—the population probably hovered between 5,000 and 20,000 for about a millennium—it would have been a megalopolis in its day. Most people living in the region at that time had never seen a settlement larger than a village of about 200 people. Built from mud and thatch, Çatalhöyük was a vast warren of interconnected dwellings whose interiors were accessed via ladders and rooftop doorways. Though its inhabitants had no writing, they left behind thousands of figurines, paintings, and symbolically decorated skulls.

Sometime in the mid-sixth millennium BCE, the people of Çatalhöyük left its busy, cramped sidewalks behind. There were many reasons: drought in the Levant region, problems with social organization, and possibly the layout of the city itself. Most people who left did not found new kinds of cities; instead, they returned to village life or nomadism. It was as if they were rejecting not just Çatalhöyük, but city life itself. Over time, the city and its roads were buried in layers of sand. By the time European archaeologists "found" the city in the 20th century, its culture was known to the local population largely as a myth. Turkish farmers were aware that an actual city lurked beneath the hills because their plows turned up exquisite artifacts on a regular basis, and a few walls still poked up out of one hilltop. But nobody knew much about who had lived there.

Something about Çatalhöyük *has* been lost, even if locals always knew where it was. Researchers are still struggling to understand what the people of Çatalhöyük believed about their world. When I visited, archaeologists were in a heated debate over whether the people living there had concepts of history or spirituality—or both. Why did they paint specific ochre designs on the walls of their homes? Why did they decorate their doorways with bull horns? Why did they bury their dead beneath their beds? We have some ideas, but nothing is certain. We've lost the cultural context that made it meaningful to people who called it home thousands of years ago. Still, its residents left enough behind that we can reconstruct what their everyday lives were like, along with the problems that made urban life more trouble than it was worth.

The next city we'll explore was never forgotten, though its exact location appears to have gone missing for a time. Pompeii, a Roman tourist town on the sunny shores of the Mediterranean, was buried deep under volcanic ash after the eruption of Mount Vesuvius in 79 CE. Eyewitnesses and historians recorded the city's horrific fate, but Pompeii wasn't systematically excavated until the 18th century.

It would seem that there's a rather simple cause for Pompeii's abandonment. Nothing like 482°C pyroclastic flows blowing through town to clear everyone out. But that's not the whole story. Pompeii had weathered natural disasters in the past, recovering from extensive damage it suffered in an earthquake over a decade before Vesuvius erupted. People who lived there knew it was a dangerous place. Indeed, over half the residents evacuated on the morning of the eruption; they fled when the mountain started smoking and triggering quakes several hours before the deadly blast.

Popular accounts of the city's demise suggest that Romans shunned the buried city out of superstition and fear, quickly losing track of where it had once stood. Nothing could be further from the truth. Pompeii's demise was followed by one of the greatest relief efforts in ancient history. Emperor Titus toured Pompeii twice after the eruption to assess the damage, discovering that the once-lush landscape was entombed in thick, superheated ash, oozing toxic fumes. Pompeii was unsalvageable. Titus and his brother Domitian, who succeeded him, used the sprawling empire's wealth to rebuild the lives of people whose homes were lost. They allocated money to survivors, and paid workers to construct homes for them. Archaeologists have recently uncovered new evidence of the empire relocating refugees to nearby coastal towns like Naples, expanding neighborhoods and roads to accommodate them. Many patricians perished in the blast, leaving fortunes behind, so the government allowed freed slaves to inherit their masters' commercial holdings. These freedmen made prosperous new lives for themselves. Pompeii may have been lost, but Roman urbanism continued to thrive.

Thanks to the ash that encased Pompeii in 79 CE, we can get an unvarnished picture of the cosmopolitan culture that Romans worked so hard to preserve. The century leading up to Pompeii's demise was a time of great change for the empire, when women, slaves, and immigrants gained rights and penetrated the inner sanctums of political power. A new kind of polyglot public culture was emerging, and we can track its progress in Pompeii's streets, where ordinary people scrawled graffiti, got drunk in tabernas (pubs), and socialized in bathhouses and the city's infamous brothel. It continued to shape urban life in the West for millennia to come. The fate of Pompeii is evidence that the demise of a city isn't the same thing as the collapse of the culture that sustained it.

Fifteen hundred years later, Angkor suffered a slow-motion version of the catastrophe that Pompeii experienced in just one day. Instead of a single volcanic eruption, the city was pummeled by climate crisis lasting a century. The timescale may have been different, but the results were similar: environmental disasters like the floods that Evans described at the West Baray made the city unlivable for the majority of its population. But the final blow had nothing to do with nature: the Angkorian kings could no longer command armies of laborers to rebuild the canal system that was the city's lifeblood. Perhaps the most unsustainable part of Angkor's urban planning was not its system of reservoirs, but a rigid social hierarchy that depended on forced labor.

Meanwhile, in the Americas, another great medieval city expanded and contracted, its reversal of fortunes recorded indelibly on the landscape. Cahokia was the largest city in North America before the arrival of Europeans, growing from a small riverside village in the Mississippi River bottom to a sprawling metropolis of over 30,000 people whose holdings straddled two sides of the river. The Cahokians built towering earthen pyramids and elevated walkways where St. Louis, Missouri, and East St. Louis and Collinsville, Illinois, stand today. Their homes and farms were spread between ceremonial centers where festivals attracted people from all across the south. Between 900 and 1300, Cahokia was the center of "Mississippian" culture, a social and spiritual movement that united towns and villages all along the great river, from Wisconsin to Louisiana.

I spent two summers at an excavation in Cahokia where archaeologists uncovered part of a busy residential neighborhood near Cahokia's greatest ceremonial pyramid, nicknamed Monks Mound. Built entirely by people carrying baskets of clay from "borrow pits" nearby, Monks Mound is 30 meters high, and has a footprint the size of the Great Pyramid at Giza. But archaeologists Sarah Baires and Melissa Baltus weren't interested in who lived at the top of that pyramid. They wanted to know how ordinary people lived in Cahokia.

On my hands and knees in the mud, my ankles bitten by flies and my neck burned by the sun, I came face to face with what Baltus calls "deliberate abandonment." When Cahokians were finished using a structure, they would seal its fate with a ritual. They pulled up its walls of wooden poles, tossing them aside to use as firewood. They carefully filled the empty postholes with colorful clay, and sometimes with bits of broken pottery or tools from the lifetime of the house. On the floor of one structure, Baires and Balthus found a massive posthole that had been sprinkled with a layer of blood-red hematite flakes. Sometimes, the Cahokians would light a fire in what remained of the structure, burning household items with it. Once the fire had

9

died down, residents would "seal" the abandoned place with a layer of clay and build a new one on top.

Occasionally, this ritual of deliberate abandonment extended to entire neighborhoods. Archaeologists excavating in East St. Louis found a field where dozens of house effigies burned all at once, their walls eaten in a fire that also consumed offerings of corn, ceramics, and beautifully crafted projectile points. Perhaps the Cahokians believed that every built environment had a set lifespan, and always expected to close up the entire city one day. If that's the case, Cahokia may have been designed with termination in mind, its fate sealed even as the mounds rose to incredible heights.

Why would people put so much work into building a city if they knew it was going to die? Seven years ago, when I started researching this book, I had never considered that question before. I was fascinated by Çatalhöyük and Cahokia, but I stuck to modern cities in my work, trying to glimpse the future of humanity in the streets of Casablanca and Saskatoon, or Tokyo and Istanbul. I intended to write about how the cities of tomorrow would last forever, if we designed them properly. Then something happened that made me want to investigate the past.

I returned from a week of research in Copenhagen to discover that my estranged father, always an angry loner, had committed suicide. We had barely spoken in years. While I was in Denmark talking to scientists and engineers about the future of cities, he was composing a long suicide note that veered from instructions on how to care for his beloved flower garden, to his rage at losing a fight with the city to preserve a redwood growing at the edge of his property. On the phone with the coroner, I felt numb. I knew he was unhappy, but I thought he was going to get better. I hoped one day we'd have a normal relationship. Every death is hard in its own way, but the sorrow from suicide is saturated by one, painful question. Why would he choose death, when he had so many other options?

I went through my father's papers, his half-dozen unpublished novels, and his emails, looking for anything that would explain what had driven him away from me, and then away from the world entirely. There were dozens of answers, or maybe none. I asked myself why he took those pills until I couldn't stand it anymore.

Trying to focus on something completely different, I visited Çatal-höyük during excavation season. I thought maybe a journey into the deep past would help me escape my sadness in the present. When I got there, I found myself surrounded by people whose entire job is to study the ways of the dead, and who learn about ancient life from graves. You'd think that would be terrible for a person in my frame of mind, but it was exactly what I needed. Inspired by archaeology, I was finally able to stop asking why my father had killed himself. I turned to a much harder question: How did he live? What comfort could I take from the things he'd taught me, and what could I learn from the choices he made? Answering these questions was my first step toward healing.

It was also the spark that led to this book. I realized that every city's death feels like a mystery because we usually look at its demise in isolation. We focus on the moments of dramatic loss and forget its long life history, in which people spent centuries making millions of decisions about how the city would be maintained. I don't think we can understand why people choose to let their cities die until we consider the specific ways they lived as urbanites.

That means asking deceptively basic questions. Why did our ancestors leave the freedom of the open land for cramped, stinky warrens full of human waste and endless political drama? What kinds of counterintuitive decisions led them to settle down and plant farms whose crops could easily fail, leaving them to starve? How did thousands of people ever agree to live together, cheek to jowl, cooperatively building public places and resources for strangers to enjoy? Hunting for answers, I wandered through the remains of the abandoned cities in this book. I immersed myself in their life stories, and put in years of research trying to unravel just some of their cultural complexities. To understand why people fled, I needed to know why they had come, and how hard they worked to stay. I wanted to appreciate what they lost when they abandoned the homes they had built.

The histories of Çatalhöyük, Pompeii, Angkor, and Cahokia are dramatically different, but all of them span centuries of constant transformation. Their layouts changed along with their citizenry. Immigrants came to these cities from near and far, lured by everything from delicious food and specialized work, to entertainment and the chance to gain political power. Most important among those immigrants were the laboring classes, who often accounted for more than two-thirds of urban populations. Leaders rule from their mounds and villas, but a city is truly maintained by the ordinary working people who farm, run shops, and build roads. Before the industrial revolution, the most valuable economic and political power came from human labor. But that labor came in many forms. Sometimes it was domestic labor, where some people in a family were responsible for maintaining the house, tending the flock, or doing the cooking. As cities grew, elite classes organized labor by enslaving people in various ways, including indenture, or by turning them into serfs. Creating cities was in many ways about organizing labor, whether by force or by enticement. Usually it was a combination of both. And when their cities stumbled politically and environmentally, laborers felt the squeeze more than anyone else. They had to decide whether they should stay and clean up, or start again somewhere else.

Four Lost Cities is about the tragedies in humanity's past, and it's about death. But it's also about recovering from loss, by taking a clear-eyed look at where we've been and the decisions that brought us there. Today in cities across the world, we face the same problems our urban ancestors did, as politics are eroded by corruption and climate disaster looms. Because the majority of humans now live in cities, the stakes are a lot higher. The fate of urbanism is yoked to the fate of humanity. If we replicate our past failures in the 21st century, we risk spreading a form of toxic urbanism that will change the face of our whole planet—and not in a good way. Already, cities are struggling to prevent water contamination, food shortages, pandemics, and homelessness. We're barreling toward a future in which the metropolis is unlivable, but the alternatives are worse.

The urban age doesn't have to end this way. Before they were lost, Çatalhöyük, Pompeii, Angkor, and Cahokia were home to thriving civilizations whose dark futures were by no means fated. My hope is that the deep histories in this book can show us what it takes to revitalize a city and the natural environments that surround it. After all, we learn best from our mistakes.

PART ONE

◇◈◆◈◇

Çatalhöyük
THE DOORWAY

ÇATALHÖYÜK

WEST
MOUND

ÇARŞAMb.

seasonal wetlands (7000 B.C.E.)
Konya Plain (modern day)

300m

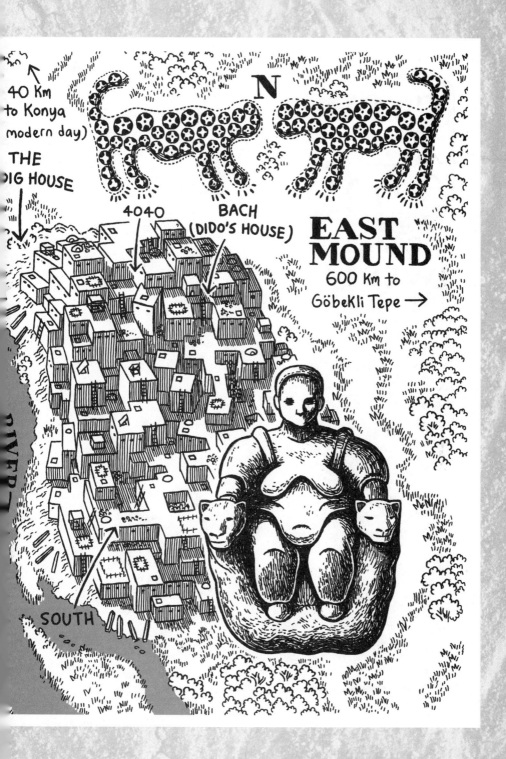

CHAPTER 1

The Shock of Settled Life

I journeyed to one of the world's oldest cities by hopping on an air-conditioned bus in Konya, a busy metropolis of two million people in central Turkey. The morning was cloudless and hot as we bounced along the road out of town, past stores selling everything from fresh eggs to Apple computers. When the gleaming apartment towers gave way to fields, we did not leave civilization behind. We passed tidy Bedouin camps by the side of the road and wound through small towns where new homes were going up on almost every block. After about 45 minutes, the bus stopped in a small gravel parking lot. Wooden cabins and long, low buildings surrounded a pleasant courtyard full of canopied picnic tables. It looked like a retreat center, or maybe a small school.

But it was actually a portal to the distant past. A few hundred meters beyond the picnic tables was Çatalhöyük, a city built before cities existed. Most of it lay buried beneath the bulk of the East Mound, a low, wind-smoothed plateau. If you looked at it from above, the 13-hectare East Mound would form a teardrop shape, its contours like an earthen blanket draped over the 9,000-year-old remains of a city whose inhabitants built houses atop houses for so long that the layers of clay brick construction formed a hill. Beyond the East Mound was the newer West Mound, a smaller neighborhood that formed about 8,500 years ago. When this city was young, rivers flowed around these city-hills, and farms were scattered across the Konya Plain nearby. Today the land is dry and covered in patches of yellowing grass. I sucked in a breath of warm, dusty air. This is where it all started. The world I knew—full of condos, factory farms, computers, and cities swarming with thousands of people—was born in places like this.

Some archaeologists call Çatalhöyük a "mega-site," or an outsized village where several smaller settlements merged. The city appears to have grown together organically, without any centralized planning or direction. Çatalhöyük's architecture is unlike anything we see in the region thereafter. Each house was constructed like a cell in a honeycomb, pressed tightly against its neighbors, with almost no streets to separate them. The city grid was at least one story above the ground, and sidewalks wound across rooftops, with front doors cut into ceilings. Residents would have spent a lot of time on their roofs, cooking and crafting tools, often sleeping outside under light shelters. Simple wooden ladders helped residents climb up to the city, and down into their homes.

When the earliest construction began, many people coming to live at Çatalhöyük were only a generation or two removed from nomadism. The idea of settling permanently in one place was revolutionary at the time. Though there were little villages before Çatalhöyük, the vast majority of humans roved in small bands, just as their Paleolithic ancestors had done for over a hundred thousand years. Imagine leaving the natural world, where your companions were only a few people and animals, and embracing sedentary existence in a box crammed next to hundreds of other people in boxes. Your parents and grandparents, who knew only the old nomadic ways, couldn't possibly have prepared you for the weird complexities of urban life. It's no surprise that the people of Çatalhöyük struggled to figure out the best way to live together, and made many deadly mistakes along the way.

This was perhaps the first time in human history when the question "Where are you from?" came to matter as much as "Who are your ancestors?" For a nomad who is always on the move, "where are you from?" is difficult to answer. What matters is who your people are. That's why many ancient texts in the West, including the Bible, introduce their heroes with interminable lists of fathers, grandfathers, great-grandfathers, and so on. You are literally the sum of your ancestors. But when you live in one city your whole life, that place can become even more important to your sense of self than your family heritage.

When people passed through one of Çatalhöyük's thousands of rooftop doorways, they entered a new phase in human society. They found themselves in an alien future where people's identities were tied to a fixed location; the land was theirs, and they were part of the land. It would have been like a slow-motion shock, reverberating across generations. Survival now hinged on whether the climate was favorable for farming, and death could come any year from drought or flood. As we'll see in the story of this ancient city, settled life was so difficult that humans nearly decided to slam the door on urbanism forever. But we didn't. And that's what led me here, millennia later, to figure out what the hell our ancestors were doing.

The opposite of Indiana Jones

I turned my attention back to the Çatalhöyük Dig House where the bus dropped me off. It had been home to hundreds of archaeologists over the past 25 years, all of whom tirelessly worked to uncover the ancient city's secrets. I'd arrived in time to join a few dozen of them at a conference on the history and religion at Çatalhöyük.

A group of us walked to the top of the East Mound, where archaeologists had skinned away the northern surface of the hill to reveal Çatalhöyük's urban grid. This dramatic excavation, known simply as 4040, is roughly the size of a modern city block. 4040 is protected by a huge shade structure that arcs over the East Mound like an airline hangar fashioned from wood and opaque white plastic. When I ventured inside, the intense sunlight was filtered to a pleasant glow and the air cooled. Before me stretched hundreds of interlocking rooms made from golden-brown mud bricks.

At least a dozen archaeologists were working in the space, stooped next to walls, taking notes on clipboards, or capturing the morning's finds on cameras. Sandbags were stacked everywhere, shoring up what remained of crumbling walls. Our little group stood about a meter above what would have been the floor of a house, looking down into someone's 9,000-year-old living room. I could see layers of plaster stuck in clumps on the thick clay walls, reminding me of the six layers of different pastel colors I'd scraped off wooden doorframes in my 100-year-old house. In several places, red ochre designs painted by residents were still visible, zigzagging across bright patches of plaster. One was a repeating pattern of diamond-shaped spirals. Another rippled with tiny rectangles that flowed between snaking lines, as if the painter wanted to evoke a river. All of these designs were abstract but elaborate, conveying a sense of motion, as if the painters wanted the never-changing settlement to feel charged with frenetic life.

Throughout the excavation area, we saw small oval pits dug into the floors: these were the telltale signs that skeletons had been removed from their graves. People at Çatalhöyük kept their dead close, right below the elevated clay platforms that served as their beds. Bodies were buried in fetal positions, so death's shape was that of a round storage vessel, rather than the Western world's familiar elongated coffin. Some bed platforms were host to a few of these graves, while others held a half dozen. One grave, we heard later, was packed with several skulls but only one skeleton.

Guiding the group was Stanford archaeologist Ian Hodder, a soft-spoken Londoner who has directed Çatalhöyük's excavations since 1993. Hodder is basically the opposite of the Hollywood adventurer Indiana Jones. He's famous for pioneering an influential school of thought called contextual archaeology,[1] which treats ancient artifacts as keys to understanding ancient cultures, rather than as loot. If Indiana Jones had been a contextual archaeologist, he would have left that golden idol from *Raiders of the Lost Ark* in its temple, and tried to understand how it fit into the belief systems of the people who built such an incredible booby-trapped monument. When Hodder finds a priceless treasure at Çatalhöyük—and he has been privy to the discovery of many—he wants to know what it can tell us about the social relationships in this ancient city.

Taking off his floppy canvas hat, Hodder climbed into a deep, perfectly square trench cut into the floor of a house. One side of the pit was what archaeologists call a profile, a cutaway view of all the layers that represent the many centuries of houses built upon houses in this area. The lowest layer is the oldest floor, and each subsequent one is newer, which is why archaeologists will often confusingly call something "upper" when they mean "more recent." Another word for this analytic technique is "stratigraphy," or the study of the earth's layers in their historical context. Hodder pointed to the upper layers in the profile, which formed a gentle wave pattern made from layers of black material sandwiched between light brown clay, topped with a layer of black, capped by another layer studded with what seemed to be chipped bone. It looked like one of those intricate Viennese layer cakes, except three meters tall and made of dirt. We were looking at what happened to houses over hundreds of years in this city, Hodder explained. The brown clay layers formed because Çatalhöyük families tended their floors carefully, often resurfacing them with plaster. And the black layers were ash, representing periods when the house was abandoned. Often an abandoned house would be symbolically "sealed off" with the ritual burning of household objects, leaving a distinctive layer of charred materials. Sometimes the house became a trash pit after that, and neighbors filled it with more ash from their hearths, along with other refuse.

Eventually, a new family would rebuild the house, slathering a thick layer of clay and plaster over the ash and re-creating the exact layout of the older structure. Hodder described housebuilding at Çatalhöyük as "repetitive"—residents didn't value the idea of changing architectural fashions. In one case, Hodder and his colleagues unearthed a house that had been rebuilt four times, with successive residents storing their cooking pots and burying their dead in the exact same places.

In the upper levels the house he was showing us, Hodder identified three clay floors sandwiched between ash, representing distinct phases of abandonment and rebuilding. Things got vaguer in the lower levels, but we could make out at least eight more layers of interleaved clay and earthy fill. Hodder speculated that these might represent many earlier houses, or fewer houses that had repaired their floors extensively during use. Either way, we were witness to an early version of an urban phenomenon that still exists in cities today. People at Çatalhöyük created new houses out of old ones, just as I myself had made a new home in my century-old house by replastering the exterior, rebuilding some of the walls, and covering them with a fresh layer of paint.

We left the 4040 shelter and Hodder led us across the top of the mound to the southwest, heading for an older excavation called South. As we passed a few canvas tents covering smaller digs along the way, I imagined Çatalhöyük residents walking this same path across town, over the city's roofs. Though the current excavations are extensive, they've uncovered only 5 percent of the city itself. Beneath our feet were thousands of homes, built atop each other for over a millennium, their treasures still unseen.

The South dig is breathtaking. Sheltered beneath a steel-and-fiberglass structure, we could see that archaeologists here had dug at least ten meters down, uncovering ever more ancient layers of the city's grid. Standing on a wooden viewing deck, I gazed at stratigraphic layers writ large. Far below, I could see the earliest parts of the city, when people first decided to settle here year-round instead of walking the path of nomadism. Back then, this land was marshy and lush. None of those settlers had any concept of what a city might be until they started building. They kept adding more and more structures to their settlement in an ad hoc way until the clay deposits became clay houses, and the clay houses became clay rooftop sidewalks, neighborhoods, and artworks. We could see more than 1,500 years of the city's history in one glance.

Hodder pointed out a flag attached to a piece of rebar in the deepest part of the dig: "That's the dairy line," he said with a cryptic half-smile. He was showing us the layer of Çatalhöyük where scientists found the first evidence that people were cooking with dairy products. Residues in clay pots tell a story of soups made richer with goat milk, and possibly cheese. Researchers Maria Saña, Carlos Tornero, and Miguel Molist have studied sheep herding in the Neolithic, and found evidence that small sheep herds were tended by families for generations,[2] bred to provide milk and meat. But the dairy line represents more than the addition of extremely tasty items to the human diet. The emergence of dairy foods changed the way humans lived, which in turn changed the lives of animals, as well as the land around human settlements. At the dairy line, we can see the traces left behind by humans who had stopped looking for their place within nature, and started changing nature to suit themselves.

How humans domesticated themselves

In 1923, the Australian archaeologist V. Gordon Childe published a book called *Man Makes Himself* that offered one of the first stories of how urban life evolved. Influenced by the Marxist idea that human civilization changes during economic revolutions, Childe coined the term "Neolithic Revolution" to describe the constellation of developments that unfolded during the occupation of Çatalhöyük. Imagining an ancient version of the Industrial Revolution, he argued that all societies inevitably passed through an intense, rapid phase of transformation when they adopted agriculture, developed symbolic communication, engaged in long-distance trade, and built high-density settlements. He explained that this set of Neolithic practices swept through the Middle East rapidly, and from there spread around the world, sowing urbanism in its wake.

For decades, anthropology students learned about the Neolithic Revolution, an abrupt cultural break when roaming bands of nomads became citified folk who paid taxes. Though many scholars used to believe in this idea, archaeologists today have gathered new data from Neolithic societies like Çatalhöyük, making the picture a lot more complicated. We've learned that the transition from nomadic life to mass urban society was very gradual, with many stops and starts over thousands of years. Also, it didn't begin in the Middle East and radiate to the world; the set of practices we dub Neolithic emerged in multiple places independently, from Southeast Asia to the Americas. There is no question that Neolithic technologies and living arrangements changed the course of our civilizations. And the transition would have been jarring sometimes, especially for individuals leaving the old ways behind. But the Industrial Revolution is not a good analogy for the social changes we see at Çatalhöyük. During the early 20th century, one generation witnessed the widespread adoption of electricity, telephones, and cars. But over 10,000 years ago, during the Neolithic, it took dozens of generations to develop farming, and dozens more before we reached the dairy line. Still, despite progressing slowly, Neolithic people did manage to transform everything in the world around them, just as their distant progeny would with their fossil fuels and carbon-belching engines.

By the time of Çatalhöyük's founding, humanity had its own distinct ecological footprint full of animals and plants we cultivated,[3] like goats, sheep, dogs, fruit trees, several forms of wheat, barley, and many other crops in different regions of the world. Along with these, we attracted life-forms we weren't expecting, like mice, crows, weevils, and other vermin—plus plague-causing microbes that could jump easily from human to human, or animal to human, in the close quarters of a settlement. The human ecosystem is a complicated web of desirable and undesirable life-forms, attracted to our food, waste, bodies, and shelters.

Humans changed every life-form that entered our settlement ecosystems. We bred plants so that their edible bits would ripen faster and feed more people, which led to wheat with bigger seeds as well as plumper fruits. Domestic animals like dogs, sheep, goats, and pigs changed during thousands of years of domestication, too. Perhaps the most obvious change is called neoteny, or the process of becoming more childlike. Domestic animals tend to be smaller, developing softer facial features like floppy ears and short snouts. Other changes are more dramatic: domestic pigs have an extra pair of rib bones. Humans weren't exempt from this process. We also domesticated ourselves.

Generations of settled life, eating a wide variety of soft, cooked foods, left their marks on our bodies. Neoteny has made human faces more delicate and thinned out our body hair. Our jaws became shorter and more rounded, which may have allowed new sounds to enter our languages.[4] Specifically, the "v" and "f" sounds, produced when our top teeth press against our lower lips, are only possible in mouths where the lower jaw bites behind the upper. And this configuration, in turn, is likely the result of eating the kinds of farm-fresh grain mashes and stews that agriculture made possible.

New kinds of foods also caused a huge segment of the human population to undergo neoteny at the genetic level. All human babies are born with the ability to digest lactose, a type of sugar found in raw milk. Before the Neolithic, we became lactose-intolerant as we aged, experiencing extreme gastric distress if we drank a glass of milk or ate cheese. But once milk products entered the human diet in the West, a genetic mutation for adult lactose tolerance spread like wildfire through the population. It was a dramatic and widespread genetic shift, and it happened entirely because of our shift to settled life. In the artificial ecosystem of the city, no life-form was unchanged—including *Homo sapiens*.

This transformation would have been obvious to people at Çatalhöyük, who were deeply aware of the distinction between wild and domesticated animals. Koç University archaeologist Rana Özbal, who studies food in the city, told me that people in Çatalhöyük preferred meals made from domestic plants and meat. Based on the chemical residues she's analyzed in storage containers, cookware, and trash pits, we know that people ate foods like milk, grain, sheep, and goat. Wild animals like aurochs were generally eaten only on special occasions, like a public feast. Domestication appears to be a self-reinforcing process: humans were drawn to domestic foods, which transformed our bodies, and over time our bodies were better suited to those foods, which made them all the more alluring.

At the same time, domestication changed more than human biology. It was also connected to striking new kinds of symbolic structures. Hodder described finding teeth from weasels and foxes, along with bear claws and boar jaws, deliberately embedded in the plaster walls of many homes at Çatalhöyük. Often, people would spread a thick layer of plaster on auroch skulls, leaving the horns intact, and mount them next to their doors. Inside many homes, these plastered heads were stacked atop each other on pillars, creating the illusion of a ribcage made from horns. Wild animals have a starring role in paintings, too, where we find leopards, aurochs, and birds. Stanford archaeologist Lynn Meskell points out that the most common figurines at Çatalhöyük are animals, not humans;[5] only a tiny fraction of the hundreds of figurines found at Çatalhöyük represent people or human body parts.

Why would a society with a taste for domestication be so fascinated with the wild world they're desperately trying to leave behind? Though humans in the city were domesticated, most were only a few degrees of separation from nomads who lived without walls, surrounded by animals who might become predator or prey at any moment. Hodder speculates that wild animals remained a source of spiritual awe for city folk, and people used their images to invoke power.[6] One of Hodder's favorite wall paintings shows two leopards standing toe-to-toe, their fierce faces turned away from each other to gaze pitilessly at the viewer. In another mural, gigantic vultures appear to be making off with people's heads. Hunting scenes feature tiny, stick-figure humans looking puny next to dramatically enlarged bulls and boar. Wild animals loomed large in people's imaginations at Çatalhöyük—often quite literally.

Humans weren't always depicted as being at odds with their wild counterparts. A favorite subject for Çatalhöyük artists was the therianthrope, or a human-animal hybrid. In one painting, a vulture has human legs. Many humans are drawn with leopards' spots while they hunt or taunt bulls. Archaeologists have found weasel and other predator scat placed deliberately into human graves, as if to combine "dirt" from a dangerous animal with grave dirt. Perhaps this was one way humans claimed symbolic power, boasting that they were as swift as leopards, as dangerous as vultures, or as bloodthirsty as weasels. Hodder suggests that wild animals might have been treated as powerful ancestor figures from the past, and having relationships with them granted authority to people in the present. Put another way, the therianthrope might have been an early form of political posturing, a way of asserting authority over other people by claiming to be a bit more than human.

But maybe the wild animals in the walls were meant to remind urbanites of a time when their ancestors slept in bedrolls or tents, flimsy structures that could not withstand an auroch attack. Looked at from this perspective, wild animal imagery was a reminder of human weakness. Our walls, now strong enough to filter out predators, were once fragile. Wilderness lurked just beyond them, waiting to claw its way through. Marc Verhoeven of the RAAP Archaeological Consultancy interprets the walls at Çatalhöyük as places for "hiding and revealing"; the untamed world is invited in, only to be plastered over. After all, domesticity doesn't mean shutting out nature. Instead, it's more a filtering process, allowing certain life-forms inside while keeping others at bay. Domesticated animals, plants, and people live inside the house, while the wilderness remains trapped in its walls. Çatalhöyük's urban design reflects a society adjusting uneasily to domesticated life. Its people held onto their wild past because it gave them power, but they wanted to keep it contained, at a remove.

There was something else that the people of this ancient city wanted to keep at a remove: their neighbors. In this regard, people living in Istanbul's gleaming high-rises would have a lot in common with their Neolithic precursors. Crammed together permanently, Çatalhöyük's people struggled to maintain privacy with only about 60 centimeters of mud brick between themselves and the people they saw every day. Anthropologist Peter J. Wilson writes in *The Domestication of the Human Species* that cities like Çatalhöyük existed at the dawn of the concept of privacy.[7] As nomads, humans had very little alone time. Space was shared, and houses were collapsible, providing courtesy screens rather than actual separation from the group. That said, absolute privacy was guaranteed to those who wanted to leave the group and go their own ways. If two groups got into a conflict they couldn't resolve, they didn't have to stew on either side of the same wall. They could simply walk in different directions.

Çatalhöyük turned this social formula on its head. People could conceal themselves completely in their homes, shielding everything they did from the eyes of their neighbors. But in a permanent settlement where people acquired a lot of possessions, leaving the group behind became extremely difficult. As a result, the doorway into someone's home became a boundary saturated with both social and mystical power. Wilson writes that when someone asks to enter a home, they request that the host "expose or reveal something of [her] private domain to neighbors."[8] Urban society is full of closed doors and hidden rooms, which gave people a new way to interact with each other, exposing only parts of themselves. Ironically, it took the invention of a city for people to conceive of being alone, away from the crowd. Put another way, the concept of privacy had arrived, and with it the concept of a public.

Back in the South excavation shelter at Çatalhöyük, I looked deep into the layers of the city: walls built on walls, floors on floors, all revealed in an enormous staircase of levels that led backward in time. I realized that this city wasn't merely a physical structure. Along with houses, its residents were building new layers to their identities. Inside their homes, they could do things that nobody knew about. No doubt sound leaked between the walls, and gossip networks broke silences, but people here had a novel psychological sense of being able to get away from other people even while staying among them. Opening the door to go outside meant putting on a public face, and with it a set of behaviors that might be quite different than the ones acceptable within the home. The public sphere existed above ground, on the rooftop sidewalks, while the private world existed on the earthen floor below. And beneath it all was the realm of buried ancestors and ritual mementos, in a space beyond public and private. In short, the house was a way to think about social relationships.

The longer people lived on one plot of land, the more that land became part of who they were. You might say that this was the earliest stirring of feelings that led to phrases like, "I'm a New Yorker," or "I'm from the prairies." These statements are meaningless unless you already associate selfhood with a fixed location. Hodder and other archaeologists describe this way of thinking as "material entanglement," when our identities become bound up in the physical objects around us. Those objects might be anything from a ritual weapon or a gift from someone we love, to a hill where we were born. At Çatalhöyük, houses are the most obvious sites of material entanglement, for spiritual and pragmatic reasons: their walls were studded with wild magic, their floors contained powerful history, and their storage rooms held enough food to keep an entire family alive without anyone having to venture beyond the safe, domestic realm of farm and flock.

Humans had the technological ability to build houses long before we started living in them full time. So it wasn't as if we had a technological breakthrough that led to a new way of thinking. Indeed, it might be the reverse. As societies became more complex, we needed more permanent objects to think about ourselves.

Laying claim to the land

Freie Universität Berlin archaeologist Marion Benz has been fascinated by this question for most of her career. She told me that settled life caused a culture shock that's still reverberating across human civilizations today. To cope with that, or perhaps to express it, people built monumental structures that converted ordinary stretches of land into fantastical landscapes. Stone monoliths, pyramids and ziggurats, and even today's mega-skyscrapers express the same impulse to tie humanity to a specific, special place.

Benz argued that we see outbreaks of monumental architecture at tipping points in human civilization, when we move from one mode of community-building to the next. We see some of the first examples of this pattern in early Neolithic architecture from thousands of years before Çatalhöyük became a city. Roughly 12,000 years ago, seminomadic peoples created an incredible structure on the summit of a high plateau known today as Gobekli Tepe. Located about 300 kilometers east of Çatalhöyük, the site is populated with over 200

T-shaped stone pillars, some towering 5.5 meters high. It's slightly reminiscent of Stonehenge, but far more elaborate. The pillars are crawling with reliefs and carvings of wild animals, many of them dangerous or poisonous.

Evidence of periodic habitation at the site—mostly refuse from feasts and campsites—suggests it may be one of the first human settlements built in the West. But nobody lived there year-round, the way they did at Çatalhöyük. Visitors to the site reached it by following a narrow path from the mountains, and probably camped next to the pillar complex. The stone pillars, quarried nearby, stood inside a series of nested, circular walls that flanked a winding pathway to a central area full of benches and two of the tallest monoliths. The structure probably had a roof, creating a dark maze where the pillars' animal imagery would have flickered in torchlight, seeming to come alive in the wriggling shadows. Archaeologists uncovered carved and painted human skulls at the site, with tiny holes drilled in their crania so they could be threaded with leather string and hung from the stones.[9]

Physically imposing and unforgettable, Gobekli Tepe was a place where people returned for thousands of years, adding to its imposing stone structures and holding ceremonies and feasts. Klaus Schmidt, the archaeologist who led the excavation there in the 2000s, believed it was a proto-temple representing a cult of the dead.[10] But for Benz, the exact purpose of the structure is less important than the fact that humans built it to be both lasting and imposing at a time when they were first moving into permanent settlements. It was, Benz said, a way that humans asserted their claim on the land, tying human community to a place rather than to a group of people.[11]

But it was also a coping mechanism to deal with a social crisis. As people left nomadic bands to form agricultural communities, their populations grew in size. Suddenly, a community wouldn't be an extended family of people whose faces you knew by heart. In a village of 200 people, or a city of thousands, even neighbors might be strangers. People needed more than personal connections to feel part of the group. "[They] needed huge monumental art to create commitment and remind people constantly of their collective identity," Benz told me. You might say that people went from identifying with each other to identifying with a special, shared location. Symbolic landscapes replaced the nomadic tribe, both literally and emotionally.

By the time people settled at Çatalhöyük, 2,000 years after the creation of Gobekli Tepe, there had been a dramatic shift in the way people viewed their relationship with place. During that two-thousand-year gap, settled communities spread throughout the Middle East, and the shock of agricultural life faded. Benz explained the signs of this shift by tracing how animals are represented in art during this period. At Gobekli Tepe and sites of a similar age, there are some human figures, but they are "surrounded by a panoply of wild animals." Artists show us a world where humanity and wild animals are equivalent to one another. Occasionally, at Gobekli, the animals seem to overwhelm human figures. Some of the T-shaped stones have arms and loincloths carved into their lower halves, but no faces. Instead, their upper bodies are covered in animals and abstract designs. But at Çatalhöyük, there are wall paintings where animals are surrounded by human figures bearing weapons. "We see a group of hunters . . . [and] together they are successful in killing the wild animal," she explained. Benz sees "massive conceptual change" in this shift. The people of Gobekli were struggling to consolidate new societies in the wilderness, while the people at Çatalhöyük were part of "established, self-confident communities" numbering in the thousands.

Gobekli Tepe's monumental images of wild animals and public

displays of painted skulls exist on a smaller scale inside people's houses at Çatalhöyük. In the city, they became private, domestic objects, associated with hearth and home. This could be a sign that the people of Çatalhöyük no longer felt an urgent need to establish identification with a single place. People in the city were so entangled with their material environment that they could walk for several blocks without ever touching ground that wasn't formed by human hands. No one at Çatalhöyük would ever question whether humans could change their environment, and thrive in a structure that far outstripped anything the nomadic world had seen. Benz speculated that this might explain why Çatalhöyük's architecture is "anti-monumental." There are no great houses or towering monoliths. Instead there is simply the awe-inspiring sprawl of the city itself, thousands of interlocking homes, the tamed fields around them growing generation by generation. Çatalhöyük was always a place in transition, a doorway into the urban future but also a monument to the wild, nomadic past.

Getting abstract

As Çatalhöyük grew older, its residents adjusted to mass society by creating networks of trusted people within the city—others who shared their beliefs, or their skills. With its population of thousands, the city was big enough that these networks might include strangers, so people needed a quick, easy way to identify themselves, along with their affiliations. That's why the people of Çatalhöyük and nearby settlements began to carry small, decorated clay tokens that archaeologists call stamps. Usually a stamp was roughly the size of a business card, with an image in relief on one side. There's evidence that some were worn as pendants, while others were traded. Some were used as actual stamps, dipped in paint for use with textiles or pressed into soft clay to create a pattern.

Early stamps are covered with Neolithic imagery that's now familiar to us: vultures, leopards, aurochs, snakes, and other wild animals. Others show houses, sometimes two stories high, with a simple triangular roof. Middle East Technical University archaeologist Çigdem Atakuman, who has studied stamps throughout the region, believes these tokens were like portable versions of the house, using symbolism to tie a person to a place or group. People in a particular family, or from a particular village, might all carry a stamp with the same symbol. Coming-of-age ceremonies might involve granting new adults a special stamp to mark their transition. Stamps could show membership in a group of farmers or shamans, or some other group. We don't know all the ways they were used, but they appear in settlements across the region, and some of them are found hundreds of kilometers from where they were made. They brought the symbolism of sedentary life back onto the road again.

Over hundreds of years, stamp designs become more abstract. Atakuman pointed in particular to the evolution of phallic imagery. Erect phalluses are a recurring theme in the wild animal imagery at Çatalhöyük, Gobekli Tepe, and countless other Neolithic sites. In the wall paintings of people hunting animals at Çatalhöyük, the bulls and pigs often have erections. There are disembodied erections carved into the stones at Gobekli Tepe, as well as ones that are attached to vaguely human figures. Some of the figurines retrieved from Çatalhöyük appear to be disembodied phalluses, and we see those again on stamps. These phalluses have aroused a lot of debates among archaeologists. Do they represent male power? Fertility? Excitement and violence? As we'll see when we explore phallic imagery in other cities, a phallus is not always a penis. It's a symbol for a wide range of things, many of them unrelated to sex and gender. And the fate of phallic imagery on stamps tells a story of people whose society is entering a new phase.

Eventually, Atakuman explained, phalluses on stamps got more abstract: early stamps portray an undeniably phallic shaft atop two oval testicles, but as the decades passed, stamps show a pointed bulbous shape atop a circle, and then centuries later a simple triangle. Those once-phallic triangles also found their way into abstract representations of houses. Anthropologist Janet Carston has observed that early city dwellers drew a spiritual connection between human bodies and houses,[12] so it makes symbolic sense that a body part would eventually morph into a house part. But that doesn't explain why people created symbols that were increasingly abstract.[13] Atakuman suggested it was a sign that people communicated with symbols so often that they developed a shorthand. They could recognize meaning in pictures that no longer looked anything like what they represented.

There's no evidence that the people of Çatalhöyük invented written language, but in their stamps we can see they were right on the cusp. Writing is a continuation of the abstraction process we see in Neolithic phallus stamps. People were identifying themselves by referring to layers of abstraction. People using the triangle might not have even realized that the shape came originally from a phallus. Instead, it was simply the roof of a house, symbolizing connection to a specific place. Or it was embedded in a larger set of unique symbols that told people about its bearer's identity. It might reveal her hometown, her trade, or prove that she had crossed over into adulthood.

As the population of Çatalhöyük grew from hundreds to thousands, its people had to get used to a lot more than domestication. They lived in a bubble of human culture, where people's kinship ties, skills, and belief systems were deeply complex and varied. During the early Neolithic, people might identify themselves as being from a family that lived in a specific place. But a person at Çatalhöyük could be descended from a revered, shared ancestor represented by an animal; they often lived in houses with people who weren't blood relatives; and they could spend most of the day making stone tools while other people brought food in from the fields to cook. Identity was fungible and intersectional. It's no wonder that urbanites carried stamps around to explain themselves and show their allegiances.

Over time, an even more complex symbolic pattern began to emerge out of the built environment at Çatalhöyük. While I was visiting the site, the State University of New York at Buffalo anthropologist Peter Biehl suggested that the act of rebuilding a house repeatedly is the first step toward creating an idea of history. Perhaps, he mused, the people of Çatalhöyük belonged to one of the first civilizations to move beyond memory into historical thinking. History, he said, is an "externalization of memory" that lasts beyond one lifetime. Perhaps people with a strong sense of place were primed to develop that kind of cognitive framework.

Harvard University anthropologist Ofer Bar-Yosef pointed out that Biehl could just as easily be describing the birth of cosmology, which he's also seen emerging from cave dwellings filled with symbolism during the Upper Paleolithic, thousands of years earlier. Çatalhöyük's people might have threaded their ancient city with bones to mark their spiritual place in the world. Bar-Josef mused that there's probably no way to disentangle history and cosmology in the Neolithic world. Both are abstract concepts that explain human relationships in the context of something larger. We have to imagine Neolithic urban culture as one where little distinction was made between the past and the spiritual realm, or between magic and science.

Hodder believes that the city begins and ends with the small acts of many people, who imbue their houses with "increased practical and symbolic importance." Urbanism at Çatalhöyük wasn't some grand plan forged by kings and warlords. Instead it emerged from an ever-expanding sprawl of houses, where humans developed the crafts, tools, and symbolism that still make cities so appealing despite their many drawbacks. "It is in the distributed processes of daily life that small acts come to have large consequences," Hodder writes.[14] What he means is that the awe-inspiring cities we know today began as expressions of humble domesticity. The city's social relationships also emerged from domesticity, alongside new ideas about community, history, and our spiritual connections to the wild past.

CHAPTER 2

The Truth about Goddesses

Sometime in the middle of the eighth millennium BCE, a Çatalhöyük woman stepped through the door to her home and fell. She landed hard on her left side, fracturing several ribs. Her chest must have ached after healing, because for the rest of her life she favored her right side for lifting, carrying, and working. As she aged, those repetitive tasks left their marks on her bones as clearly as her fall had, dramatically wearing down her right hip joint, and leaving signs of strain in her ankle and toe joints. When UC Berkeley archaeologist Ruth Tringham unearthed this woman's skeleton, she was the first person to see it in millennia. Brushing sand from empty eye sockets and a gaping jaw in the dry summer heat, Tringham suddenly recalled the words sung by Dido in Henry Purcell's opera *Dido and Aeneas*: "Remember me but forget my fate." She named the woman she'd found Dido. For the next seven years, Tringham spent her summers excavating Dido's home, trying to get to know a woman separated from her by roughly 350 generations.

On a mild afternoon in San Francisco, I met Tringham at a café known for its Portuguese pastries and excellent coffee. Though she's lived in California for a while, she still has the touch of a British accent from her youth spent in England and Scotland. Tringham has the athletic look of somebody who is ready to dig trenches in a remote part of Eastern Europe or Turkey, regions where she's spent most of her career trying to understand the ancient masses one life at a time.

When Tringham found Dido, the ancient woman was dubbed Skeleton 8115 from Building 3 in the northern part of the 4040 excavation. Tringham's goal was to turn those anonymizing number designations into something rich and personal. "I to try to look at lives of individual households when I'm excavating, because history is not a big flow from the top down," she told me with the quirk of a smile. "You have to look from the bottom up, and combine small stories, small pieces of evidence, to see a history which is dynamic."

Tringham creates videos and writes speculative stories to bring Neolithic people to life, recording both the process of excavation and the acts of interpretation that turn bones back into human beings. Like Hodder, she's interested in the context of her finds. She wants to know what Dido would have felt, smelled, and seen in her everyday life keeping house. At Çatalhöyük, focusing on a single home can also reveal a lot about the city as a whole because the Neolithic's cutting-edge technologies were largely centered on domesticity: building homes from bricks, cooking, and crafting tools and art. Tringham believes that imagining the life of a woman like Dido can give us insights into what drew people to Çatalhöyük—and, perhaps, why they left.

What would the city have looked like through Neolithic eyes? Located on an alluvial plain shadowed by distant mountains, Çatalhöyük's mud-brick skyline in the mid-7000s BCE dominated the tops of two low hills divided by a winding river. Smoke rose from hundreds of rooftops in a fragrant haze, drifting over the tiny plots of farmland that surrounded its walls. When a house stood empty for a long time, neighbors turned it into a trash heap, filling it to the brim with broken pots, gnawed animal bones, ashes, and dung, before sealing it up with clay. To re-create the urban landscape around Dido's home, we have to picture it riddled with crumbling houses under repair, as well as open pits of reeking garbage. As archaeologist Kamilla Pawłowska writes rather understatedly, "it is likely that what we would consider rather bad smells dominated at Neolithic Catalhöyük."[1]

For a Neolithic visitor, however, the scents would have been unremarkable. The most astonishing part of this scene would have been the people. Thousands of people, far more than most humans saw in a lifetime, were living together in one, seemingly endless village. It was a precarious arrangement. Deadly rifts could open easily between neighbors, as well as in the soft brick of its walls.

Dido lived in a house made from sunbaked clay and wooden beams, its interior walls plastered white and painted with abstract designs in red ochre. When she was born, the house was already at least 40 years old, and the city itself was roughly 600 years older. Like a modern woman living in New York or Istanbul, Dido might sometimes have mused on the generations who raised children here before her. But most days would have been devoted to more ordinary concerns. After tending hearth in the morning, she climbed a ladder, opened a door in the ceiling, and then emerged into an environment created entirely by human hands. To fetch food and water, Dido strolled between rooftop workshops, goat pens, shade canopies, and small braziers for cooking outdoors. People lived on their roofs seasonally, so she might have seen bedrolls and dinner bowls tucked away for evening use. Eventually she lowered herself down another ladder to leave the city and head down a gentle slope to a river that flowed below. On her way, she passed through a patchwork of small farms dotting the swampy landscape.

She might have seen people tending herds of sheep, or digging up fresh clay by the river to make cooking pots and bricks.[2]

When she returned to her house, burdened with water, grains, sheep's milk, fruit, or nuts, Dido had to climb awkwardly up and down the same ladders she'd mounted that morning. Tringham speculates that's when she took her fall, landing with a painful crunch on her left side next to the hearth. But, she cautions, that's just one possible interpretation. In a brief fictional story, Tringham imagines how Dido had "celebrated mightily under the moon" for her daughter's birthday and fell from her family's rooftop to the ground below.[3] Either way, Dido survived the cracked bones and lived to see her midforties, an extremely old age during the Neolithic period.

Tringham reconstructed major events from Dido's life by examining a feature of her house that would strike most modern city dwellers as disturbing. There are bodies buried under Dido's bed and floor. But these are not the remains from an ancient scene of hidden murders. As I saw for myself when I looked into those oval grave pits at 4040, Dido's people did not treat skeletal remains as something taboo or unclean. They interred their loved ones directly beneath their houses. In Dido's house, two infants and a toddler were buried together near the hearth on the south side of the room. On the north side, three adults and one child found their final resting places beneath two elevated, white-plastered bed platforms that were once piled with furs and rugs. A few other bodies were found in a side room. Dido herself was one of the last to be buried under the platforms, interred enigmatically with a woven reed basket normally seen only in the burials of children. Her bones revealed a long life of labor, and a soot-colored residue in her chest cavity suggests she had black lung disease from cooking at a hearth in a poorly ventilated room. A mature man was later buried in a platform beside hers, and last of all a young child was buried in Dido's platform. Based on these remains, Tringham speculates that Dido had a number of children who died very young, then a son and daughter who lived into young adulthood. The older man was likely the father of her children, and the child buried last may have been a grandchild or other kin. It seems that Dido lived long enough to see many of her children die, which gives her life a melancholy cast.

Dido, like her neighbors, interacted a lot with human bones, routinely digging them up and entombing them again many years later in what are called "secondary burials." Wall niches in Çatalhöyük homes were used as display cases for human skulls. Each one was lovingly preserved under plaster and paint that re-created the long-lost faces of ancestors or revered elders. Scientists analyzing the wear and tear on these skulls believe that people passed them from house to house, perhaps swapping them for different skulls. They reburied them decades later, with the remains of people who were not their kin.[4] So when we contemplate the bodies in Dido's house, we have to think about this cultural context. Some bones may not have come from her immediate family. Plus, her house was abandoned shortly after she died, so Dido's survivors may have set up house elsewhere. For all we know, Dido was the matriarch of a proud line of people whose children lived on for generations, tending the same farmland and flocks that Dido did.

Tringham and her colleagues also found commemorative items from rituals buried in the floor. At some point, Dido and her family dug a hole near the bed platforms and filled it with two boar jaws and neck bones from three different sheep—which all showed telltale signs of having been cooked and eaten—as well as shell beads and a bird beak. This wasn't trash. For archaeologists familiar with the site,[5] this collection looks like the prized remnants of a feast for many people, plus some ceremonial jewelry. Maybe these were from a celebration for someone who lived in the house, marking a birth or major turning point. At another time, someone had buried red deer bones in the floor, and bricked a half-burned antler into one wall. Like many other people in the city, Dido also had two auroch skulls covered in plaster, their snouts and horns painted a deep red.

During Tringham's excavation of the house, she and colleagues uncovered 141 clay figurines, far more than archaeologists usually see in a single dwelling. Most depicted animals. But there were also a few voluptuous female figures, their facial features smoothed into anonymity, arms akimbo with hands cupping their breasts. These clay women, sometimes called goddesses or fertility symbols, pop up throughout Çatalhöyük for its entire thousand-year history. Similar female figurines have been found at other sites in the region, suggesting they're part of a belief system that spread far beyond this city's walls. They are Çatalhöyük's most iconic symbol—and also the basis for one of the most widespread pseudoscientific myths about the place.

Sometimes a naked woman isn't a naked woman

It all started back in the early 1960s, when the British archaeologist James Mellaart was the first European to get permission to excavate Çatalhöyük. At the time, the place was known to locals as two picturesque mounds whose grassy tops still showed the faint, angular ridges of an ancient city's walls.[6] When Mellaart and his team visited, they talked to local farmers whose plows had unearthed pottery and other artifacts that suggested Neolithic craftsmanship. Excited and not sure what to expect, Mellaart cut deeply into the eastern mound in 1961, roughly 200 meters south of where Dido's house once stood. Among many other artifacts, he found a few female figurines. He was especially impressed by one of them, who was seated in a chair with her hands on the heads of two leopards. He decided she must be on a throne, and that an abstract bulge between her ankles was a recently birthed child. Further excavation revealed the figurine had come from an elaborately decorated room that he dubbed a temple. Based on this scant evidence, Mellaart announced that the people of Çatalhöyük were a matriarchy that worshipped a fertility goddess.

This misinterpretation wasn't just the product of one man's over-active imagination. Mellaart probably took inspiration from the late Victorian anthropologist James George Frazer, author of *The Golden Bough*, who hinted that pre-Christian societies may have worshipped a mother goddess. Robert Graves built on Frazer's work in the 1940s with a wildly popular book called *The White Goddess*, which argued that European and Middle Eastern mythologies all came from a primal cult devoted to a goddess who governed birth, love, and death. Graves' work electrified anthropologists and the general public. As a result, people of Mellaart's generation were primed to see ancient civilizations through the lens of goddess worship. Few scholars questioned his interpretation. Meanwhile, celebrated urban historians Lewis Mumford and Jane Jacobs were quick to embrace the idea that Mellaart had finally discovered the remains of a civilization that thrived in a time before humans had rejected female power.

Mellaart went far beyond Frazer's and Graves' claims about goddess worship by suggesting Çatalhöyük was an ancient matriarchy where women ruled over men. And that claim had to do with Mellaart's ideas about sex. There was something about the imposing nudes he'd discovered that struck him as odd: none of them seemed to have genitals. Instead, their bodies were thick and strong, flanked by fierce animals. They were the opposite of the soft, eroticized centerfold models in *Playboy*, an iconic "gentleman's magazine" that Mellaart certainly would have encountered in the 1950s and '60s. Mellaart decided that a male-dominated society would never produce female figures like the ones he'd found because they didn't cater to "male impulse and desire."[7] Only a matriarchy could produce nonsexual figurines of naked women, he concluded.

Mellaart's largely unfounded hypothesis went viral when his findings were published in the US magazine *Archaeology*, complete with several pages of lavish photographs. The *Daily Telegraph* and *Illustrated London News* also covered his finds enthusiastically. The previously unknown site in Anatolia became a popular sensation, helped by dramatic pictures of the "lost city" whose residents were so strange that women had ruled over men! Since then, Mellaart's unfounded claim about goddess worship has persisted for decades. It's often the only thing that people know about Çatalhöyük. The idea of a lost goddess-worshipping civilization in central Turkey has even found its way into new-age beliefs and inspirational videos on YouTube.

Today in the archaeology community, Mellaart's ideas are received with extreme skepticism. Though he deserves plenty of credit for identifying Çatalhöyük as a rich archaeological resource, his interpretations of its culture are contradicted by loads of evidence that researchers have discovered since the 1980s.

If Çatalhöyük wasn't a matriarchy of goddess-worshippers, then how should we interpret those female figures? Lynn Meskell, a Stanford archaeologist who has analyzed Çatalhöyük figurines across the site, believes that Mellaart and his contemporaries misinterpreted them partly because they didn't have the context provided by looking at the site in its entirety. Now that we have data from 25 years of continuous excavation, it turns out that these female figurines tell a more complicated story. First of all, women and human figures generally represent a small number of figurines compared to animals and body parts. At Dido's house, for example, archaeologist Carolyn Nakamura[8] counted 141 figurines, and of these 54 were animal figurines while only 5 were fully human ones. An additional 23 represented human body parts, like hands. Other houses in the city show a similar ratio, with animals a far more popular subject than humans of all types. If any type of symbol held sway over this community, it was more likely to be a leopard than a woman.

The other thing that Mellaart got wrong about the importance of female figurines was how they were used in everyday life. Molded quickly from local clay, baked dry in the sun or lightly fired, they were clearly not put on a shelf to be admired or worshipped.[9] Worn down and chipped from frequent handling, these figurines look like they might have been carried around in pockets or bags. Archaeologists usually find them in trash piles or jammed between the walls of two buildings. Occasionally they're buried in the floor, much like those memento bones and shells in Dido's house. It's hard to imagine people treating objects of worship so casually, tossing them out rather than placing them reverentially in wall displays the way they did their ancestors' skulls.

Meskell muses that these figurines "may have operated not in some separate sphere of 'religion' . . . but, rather, in the practice and negotiation of everyday life."[10] Dido's people may not have had a notion of religion as we know it, and thus would not have worshipped a "fertility goddess." Instead, Dido might have engaged in small, everyday spiritual acts similar to those we see in animism, where spirits reside in all things rather than a handful of powerful deities. The figurines themselves may not have been objects of reverence, but the act of creating it could have been a magic ritual. Seeking guidance or good fortune, Dido would quickly mold one from the clay next to the field where she harvested wheat. Once it was dry, she could have used it in a ritual that drained its power away. Afterward, she'd throw the clay figure off her roof along with waste from yesterday's meal. If people at Çatalhöyük used the female figures like this, it's clear why people threw them away so often. Making them was more important than keeping them.

Another possibility is that these figures represented revered village elders, women who reached the age Dido had by the time she died. Meskell points out that no two figures are exactly alike, and most have sagging breasts and bellies that suggest age rather than fertility. Perhaps when Dido and her neighbors made these figures, they were calling on the power of specific female ancestors rather than some abstract magical force. Some activities or events in Dido's culture may have required the aid of a powerful woman. Still, this practice doesn't suggest a matriarchy. We know the plastered human skulls at Çatalhöyük, revered and passed from hand to hand, came from men and women in roughly equal numbers.[11] It doesn't appear that one gender was privileged over the other, at least if we consider the way skulls were preserved.

UC Berkeley archaeologist Rosemary Joyce, who revolutionized the field with her work on gender in early societies, argues that we can't be sure female figurines would have been regarded as representing women as a group. She writes:

> Even a figurine with abundant detail that allows us today to say "this is an image of a woman" might have been identified originally as an image of a specific person, living or dead, or as the personification of an abstract concept—like the representation of Liberty as a woman—or even as a representation of a category of people, such as elders or youths, unified by some feature we overlook today when we divide images by the sexual features that are so important in modern identity.[12]

Joyce points out that it's easy to project our modern understanding of gender onto ancient peoples—which means we are always looking for ways that one gender might have dominated the other. That's exactly what Mellaart did. Instead, we have to be open to the possibility that the people of Çatalhöyük divided their social world up using other categories, like young and old, farmer and toolmaker, wild and domestic, or human and nonhuman animal.

Domestic technologies

Women's real-life jobs at Çatalhöyük were hardly magical. Based on material evidence from the city, as well as comparisons with other traditional societies, anthropologists believe that women were in charge of farming and domestic work, while men hunted and crafted. Obviously there was a lot of overlap between these two spheres of work, and some people would have defied their given roles. What's certain is that neither form of labor was easy. When Wendy Matthews conducted a study of microlayers of plaster on the floors in Dido's home,[13] she was able to measure quantities of dust that accumulated on the floor, as well as typical patterns of tidying up. At Dido's home and elsewhere in the city, she found that residents kept their houses swept clean, raked ashes out of their hearth stoves regularly, and painted the walls and floor with fresh white plaster almost monthly. In some houses, plastering also meant replicating elaborate ochre wall art with each layer. Çatalhöyük's denizens replastered so often that archaeologists have found up to 100 layers of paint on interior walls.[14] This deep cleaning must have been necessary because the only chimney was through the door in the ceiling. Dido's dung-and-wood-fueled fire probably coated the walls in soot—and gave her black lung, to boot.

But all this maintenance wasn't just the Neolithic version of *Tidying Up with Marie Kondo*. It was also a way of designating specialized spaces in cramped quarters. Çatalhöyük residents created at least two different kinds of plaster to use on the floor: a bright white, lime-based mix covered the north side of the house with its bed and burial platforms, while a reddish-brown mix covered the south side devoted to the hearth, toolmaking, weaving, and other household jobs. We see this north/south pattern replicated everywhere in the city for hundreds of years, leading some archaeologists to surmise that residents divided their homes into "dirty" and "clean" areas. The dirty area was for the hearth, with its smoke, ash, and refuse; the clean area was for the bed platforms and adult burials.

Domestic labor was a combination of craft and engineering; everything in the house had to be handmade. Floors were completely covered in soft, woven reed mats, whose complex patterns left distinct impressions in the plaster that archaeologists can see thousands of years later. Based on the number of mouse bones that archaeologists find in homes at Çatalhöyük, we know they had a major pest problem, so Dido's family would have woven reed containers to keep hungry rodents out of their grain. They also made nets and clothing, which meant starting by making a textile from leather or plant materials, and then carving bones into needles to sew it. Sheep rib bones became paddles, which they used to smooth the clay for their cooking pots. They had rooftop workshops for making knives and projectile points from both flint and obsidian; they carved fish hooks; they built bricked-in stone hearths to cook what they had hunted and gathered. And by the way, they made the bricks, too. Every aspect of homemaking would have required this family to be experts in a wide range of areas.

Because tools were so difficult to make, Dido's family also did a lot of recycling. Throughout the city, archaeologists have found knives and axes that had been extensively repaired, their blades sharpened repeatedly. In Dido's house, many bone tools had been reformed and repurposed after breaking. And of course the family constantly needed to repair their house, not to mention the oven. They rebuilt their clay oven at least twice, moving it to a different location in the process. Even animal dung was reused as fuel. Çatalhöyük was made almost entirely of materials that people today would call sustainable. In fact, the city was founded in a marshy area that was rich with soft clay, ideal for brick-making, plaster production—and fashioning a quick figurine for whatever spells you might need to cast that day.

Perhaps the highest form of clay-based technology developed at Çatalhöyük was used for cooking. When the city was founded, people cooked by putting heated clay balls into baskets with their soups and stews. This was a labor-intensive process in which the chef had to keep pulling out the colder balls and replacing them with piping-hot ones fresh from the hearth fire. By the time Dido's family lived in the city, however, artisans had developed a tempered clay that was ideal for thin, strong cooking pots. Now cooking was as simple as putting a pot over the fire and letting it simmer. No more juggling pot-boiler rocks. As Rana Özbal told me, "When your cooking technology changes—it's like suddenly having a car. It changes social relations. Less people are going to be involved in the cooking process. Transferring hot stones would be labor intensive. It's not easy. If you can put your pot on the stove, you can be doing other things while that's happening." That innovation left time in the day for work that was more creative, like making wall paintings or carving bone beads. It also gave people an opportunity to specialize in different tasks, like learning to make several varieties of plaster.

Assuming that everyone in Dido's family was able to do a few more tasks thanks to those high-tech cooking pots, even more elaborate kinds of crafts became possible. One person could be freed up to go on a two-day journey to the mountains to gather obsidian at the nearest quarry. Obsidian was akin to a luxury good during the Neolithic, prized for its strong, sharp cutting edges and reflective surface. Getting more obsidian meant more raw material for skillful stone knappers, who fashioned knives by carefully striking one stone with another, flaking off bits until a blade emerged.

Farmers also needed time to perfect their agricultural expertise. Çatalhöyük's two mounds rose out of a swamp, and the river between them flooded seasonally. During rainy season in the summer, the city would have looked like an island in the middle of a muddy landscape pocked with pools. That meant the city's farmers had to plant crops at some distance from the city, on land that stayed dry. Özbal said that there's no doubt the people of the city were accomplished farmers; we have ample evidence of domesticated wheat and other grains in storerooms, plus residues of milk from animals left in cooking pots. But, she admitted, it's hard to say *where* people were plowing the land and keeping livestock. She and other researchers speculate it could have been in foothills nearby, a lengthy walk from the city. Farmers probably spent part of the year away from home to care for the fields, perhaps working in shifts. Shepherds and goatherds would have roamed far from the city, too.

Farming is another example of what becomes possible when a large population comes together. Dido's family could spare a few members to work their farmland and tend animals for extended periods of time because they had enough people back home, with decent ceramic technology. Farming sustained the city, and the city enabled farming. This reciprocal relationship is where urbanism was born; Dido would have viewed agriculture as part of city life, not in opposition to it. Urbanism emerges roughly as the same time as agriculture.[15] Farms are essentially a specialist approach to the gathering of plants done by nomads.

The benefits of specialization might be one reason why Neolithic people flocked to Çatalhöyük, despite the fact that city life was labor intensive and quite different from what most people were used to. Villagers might have been drawn to the idea of a society that was full of expert craftspeople. A practiced brick-maker produces bricks that last longer; a dedicated goatherd means your flock can be bigger. Someone who makes figurines of animals all the time can produce detailed, creative leopard icons. There probably weren't any ex-nomads arriving in the city with dreams of becoming celebrity flint knappers. But urbanization did mean everyone's house could be packed with complex tools, along with foods that would have been hard to come by in a village of 100 people. As Çatalhöyük grew, it may have attracted new residents with the promise of abundance: more people meant higher-quality items to share and swap.

Though it was full of tradespeople, Çatalhöyük was not some kind of proto-capitalist society. Residents certainly exchanged many valued objects, but the city's relative abundance didn't lead to the production of surplus goods for some at the expense of others. Çatalhöyük's settlers had not yet invented money, and there's no evidence that certain families had vastly more property than others. Most houses were one or two rooms, roughly the size of Dido's. People at Çatalhöyük didn't have the space to accumulate more possessions than they needed—they simply stored enough to survive. "Abundance" meant having enough food to ward off total starvation, and shelter that was relatively stable. At Çatalhöyük, there was no opportunity to become rich, at least in the way we understand it today.

Beyond a sense of security, the city also promised cultural enrichment, a form of wealth that we can only measure indirectly in the archaeological record. We see hints of its power in the sheer ubiquity of wall art at Çatalhöyük; every house is full of drawings, figurines, and handmade furniture. But we can also extrapolate cultural complexity from the sheer size of the city's footprint, where thousands of people interacted with one another every day. Someone who was skilled at making figurines in a nomadic group might never meet another person with a comparable ability. But in Çatalhöyük there would be several, comparing notes and swapping stories, developing more advanced techniques by working together. In short, the city allowed people to form attachments that went beyond family. City dwellers had opportunities to socialize with other people who shared their interests, as well as the ones who shared their hearths.

People today are attracted to cities because they feel an affinity for subcultures or groups that don't exist in smaller communities organized mostly around families. Çatalhöyük may have drawn new residents for similar reasons. Ian Hodder describes a curious practice that hints at one way people memorialized their nonfamily bonds. Many of Dido's neighbors built what he calls "history houses." These houses had a larger-than-average number of plastered bull heads, paintings, and bones in their walls. And these dwellings were carefully rebuilt with the exact same dimensions many times over centuries.[16] People would even exhume and rebury the skeletons that had been placed in the floor of the earlier house, and archaeologists sometimes find dozens of skeletons in the floor of a history house. Like museums or libraries, history houses were places where the people of Çatalhöyük maintained a shared repository of cultural memories.

But they were also physical embodiments of the kinds of groups that people formed after meeting on the sidewalk and developing obsidian tech together—or hanging out down by the river, selecting the right clay for bricks. History houses might have grown out of shared interests—whether in flint knapping or something more spiritual—and a connection to the city itself. For new arrivals to the city, especially those without kin, these kinds of social groups would have made it possible to stay. They probably helped the city keep growing, too. People might move to Çatalhöyük because they'd heard it was a place for communities similar to what psychologists today call chosen families.[17]

History houses also represented an abstract idea of community that could encompass people who were unknown or absent: newcomers, strangers from another part of town, the dead. Individuals affiliated with a history house didn't have to list their biological kin as ancestors; they could look to the skulls in the walls of the house, and count themselves part of its lineage. It was a philosophical leap for people whose idea of community may have come from knowing the faces of everyone in their nomadic tribes. History houses were like bodies that lived for generations, linking past and present city dwellers, entangling their identities with the special place that was Çatalhöyük.

Still, it's not likely that people in Çatalhöyük thought of themselves as Çatalites, the way a person in Brooklyn might call herself a New Yorker. Tringham believes the city functioned like a set of loosely connected villages, each with its own subculture. These enclaves may have been holdovers from what would have seemed like the distant past to Dido, when several villages merged to form the city. Before Çatalhöyük formed, Hodder said, the Konya Plain was scattered with many villages that suddenly disappeared, as if their populations had relocated to one mega-village. A person who walked across town might have passed through different neighborhood-like clusters, perhaps separated by open areas. Maybe their residents spoke different languages and ate different kinds of foods. Still, they shared a place in common, allowing someone like Dido to meet and befriend people who seemed very different from her own family.

Obviously we can't be certain what Dido believed and felt, but we do know that she lived in a house full of items made from such a wide variety of materials, using such disparate techniques, that they could only be the result of a diverse society with some degree of specialization. We also know that the symbolic art and designs that surrounded her every day would have reflected a belief in abstract relationships. At the same time, we have to appreciate how fundamentally bizarre city life would have been at a time when almost nobody outside Çatalhöyük was an urbanite. No doubt this gave Dido and her neighbors a vague sense of dislocation, especially because they were forming communities that had never existed before. When conflicts arose, they had few precedents to call upon in order to resolve fights or enmity between neighbors. As the city neared the end of its life, social problems spread like a stain over Çatalhöyük's homespun community fabric. Though the city's residents suffered through many misfortunes and survived, it turned out that the one hardship they could not endure was coping with each other.

CHAPTER 3

History within History

I met Ian Hodder again in early 2018, when he'd just completed 25 years of leading excavations at Çatalhöyük. Seated behind a desk in his sunny office at Stanford, he talked about what he'd learned during that time. What stood out to him the most, other than the richness of the symbolism at every layer, is the way the site represents "history within history." Çatalhöyük was constantly changing, and the city people founded 9,000 years ago was dramatically different from the one they abandoned a millennium later. "We now recognize a huge amount of change," Hodder said. He continued:

> There is a moment you might call classic Çatalhöyük, circa 6500 BCE, when the whole site was densely occupied. Everywhere we dig at that level, we find dense housing. It was also a crisis point, when the aggressively egalitarian society had strains. Something dramatic has happened, and we see a lot of burning of buildings and abandonment. After that crisis, we see ritual burning regularly for 500 years.

This "ritual burning" is not an act of violence or destruction; it's part of the house abandonment ritual that he had shown me in the stratigraphic layers of ash from the floor of a Çatalhöyük house. When people decided to leave a home behind, they often "sealed" the foundation with a ceremonial layer of clay, and then burned the remaining household objects along with a few offerings.

Hodder emphasized that Çatalhöyük was abandoned so slowly after the "crisis" in 6500 BCE that the shift would have been almost imperceptible during the lifetime of a single resident like Dido. People left the East Mound over centuries, and then the West. But as people abandoned the West Mound, he said, the empty land around Çatalhöyük came to life with new communities. "The Konya Plain fills with sites. It's as if Çatalhöyük proliferates into other settlements on the plain, and the West Mound is just one. You could see it as a population explosion," he explained. He thinks the exodus from Çatalhöyük might have represented a new kind of freedom, in which people broke away from a "closed, controlling system" on the mounds. Another possibility is that people moved in response to changing food needs. People on the Konya Plain were shifting to more intensive kinds of cereal farming and sheep herding, so perhaps they required more space around their settlements. But nobody "lost" Çatalhöyük. Even when the old city was completely empty, people still used it as a cemetery. "In a way the site was never abandoned," Hodder said. "There are huge numbers of burials up until the Byzantine and early Islamic period [in the 11th century]. People remembered it and used it." Newcastle University archaeologist Sophie Moore recently found evidence that cemeteries at Çatalhöyük were still being used regularly up until about 300 years ago.[1]

Hodder is echoing a common belief among archaeologists today, which is that terms like "lost city" and "civilizational collapse" are the wrong ones to use in a case like this.² Instead, it's more accurate to say the city underwent a transition. Indeed, there never was a time when Çatalhöyük wasn't in transition from one kind of cultural arrangement to the next. That's the difficult part about studying cities: they are not static entities that remain the same over time before suddenly disappearing into nothingness. At any given moment, they are a composite of many social groups, who likely view city life in different ways. And those social groups also change over time, altering the physical and symbolic fabric of the city to reflect their worldviews. Until they stop wanting to live together.

But even when that happened at Catalhöyük, nobody "lost" the city. On the Konya Plain, the city made from the bones of ancient ancestors continued to welcome the bones of new ancestors. The place remained special, long after people left it.

By 6000 BCE, people had inhabited Çatalhöyük continuously for over a thousand years, and nobody left on a whim. With some exceptions, cities are typically abandoned the same way Hodder describes them being originally populated. Thousands of small acts empty them out, each one a hard choice. In the modern world, psychologists say that moving is one of the hardest life changes people experience, causing feelings of isolation, loss, and depression.³ Though obviously Neolithic people didn't experience "moving" in the same way we do—carrying sofas into trucks and suffering real estate investment anxieties—the act would have exacted some of the same psychological costs. And logistically, it would have been vastly more difficult. Moving meant taking what you could carry, along with any animals you could herd. When you arrived at the new place, you'd have to build a new house and find local food and water sources. The whole process would be almost impossible to do on your own because setting up a Neolithic household was labor intensive, requiring skills that ranged from farming and cooking to textile-making and house construction.

Imagine doing all that while also trying to assimilate into a new culture. In 2011, the US Presidential Task Force on Immigration identified a number of hardships commonly faced by immigrants,[4] ranging from the difficulty of learning new languages and cultural norms, to dealing with prejudice and lack of access to resources. We tend to forget all the ways that immigration hasn't changed over the millennia. Many people leaving Çatalhöyük for other places would have dealt with language and cultural barriers, as well as potential problems when they bargained with new neighbors for access to farmland. And yet despite the certain difficulties they faced, they started to leave the city in greater numbers.

The 8.2K climate event

Hundreds of years after Dido was buried in the floor of her house, Çatalhöyük entered the final phase of its occupation. The city had stood in the same place for over a millennium, and its artificial bubble of domestication was changing from without and within. The ever-growing mound of old construction and refuse in the older parts of the city had reached new heights. Slightly before the turn of the millennium between the 6000s and 5000s BCE, people began trickling away from the East Mound, where Dido and her family had lived, and built up a smaller settlement across the river. This newer West Mound, as archaeologists call it, was a thriving community for about 300 years. As the houses on the East Mound slowly crumbled, nobody came along to claim them.

When I asked Ruth Tringham why people defected to the West Mound and beyond, she joked that maybe it was because they were sick of hiking up such a tall mound with their water and food, so they struck out for territory that was a little flatter and more accessible. Joking aside, there is a grain of truth to the idea that something happened to make the land of the East Mound less appealing. Many researchers have noticed that the slow migration to the West Mound coincides roughly with a period of rapid climate change that starts around 6200 BCE.[5] During this time, Earth was emerging from an ice age that covered Canada and the northern United States in a massive glacier called the Laurentide Ice Sheet.

There's evidence that rivers on the vast Konya Plain were shifting course and drying up. The weather was getting cooler, and there was at least one period of drought. Then, as temperatures warmed, the Laurentide Ice Sheet began to melt, creating two lakes of near-freezing water known as Agassiz and Ojibway. These lakes grew to cover a large area where Ontario and Quebec are today, trapped behind natural dams created by the retreating ice. But the ice wouldn't hold for long. Eventually the Laurentide Ice Sheet collapsed, and Agassiz and Ojibway emptied rapidly, releasing catastrophic amounts of freshwater into the ocean.

Evidence from around the world reveals that sea level rose at least 30 centimeters, and in some areas as much as four meters. More importantly, the freshwater also interfered with the ocean's "thermohaline circulation," a complex interaction between saltwater and freshwater that drives currents in what is sometimes called the ocean's conveyor belt. When the thermohaline circulation is perturbed, warm water can't traverse the globe, and most of the oceans remain chilly. This affects the weather, too. In the area around Çatalhöyük, it's likely temperatures dropped about 4°C on average, and rainfall probably slackened as well. For people who were used to living in a warm, swampy cityscape, there would have been a noticeable shift to a cooler, more arid climate. The average global temperature didn't rise again for almost 400 years.

This climate shift, referred to rather blandly by climate scientists as "the 8.2k event" because it happened 8.2 thousand years ago, has been so widely documented by scientists that it serves as a model for how climate change works. In 2003, the US Department of Defense commissioned a study of the security risks posed by climate change, and researchers cited the 8.2k event as an example of how glacier melt would affect the environment and human society.[6] If the rapid glacier melt we're seeing today[7] released an amount of icy freshwater into our seas equivalent to what flowed from Agassiz and Ojibway, temperatures in Asia, North America, and northern Europe would drop by over 5°F. Meanwhile, temperatures would increase by 4°F in areas throughout Australia, South America, and southern Africa. Droughts would follow, wreaking havoc on agriculture in Europe and North America, while winter storms and winds would intensify, especially in the Pacific. Famine, wildfires, and floods would come next.

These conditions would be devastating in the modern world, but there isn't strong agreement among archaeologists about whether the chilly, dry weather was enough to drive people away from their beloved hometown of Çatalhöyük. Ofer Bar-Yosef has studied the effect of climate change on ancient human migrations as far back as 50,000 years ago, and he's one of the scientists who think the glacier collapse would have made life impossible on the East Mound. He believes that the cooler weather would have starved whole villages, driving people out of the area entirely for 200 years.[8] Several villages he's studied in the Levant were completely abandoned during the 8.2k event, only to be rebuilt after the weather warmed again. He suggests something similar might have happened at Çatalhöyük. The move away from the East Mound, Bar-Yosef argues, is evidence that people abandoned the area for centuries, and built the West Mound upon their return.

Other scholars disagree. University of Reading archaeologist Pascal Flohr and her colleagues have tracked global responses to the 8.2k event, and they see no evidence for abandonment at Çatalhöyük during the weather changes.[9] Indeed, they consider it a triumph of Neolithic resilience that people in the city managed to stay put and change the structure of their city, rather than pulling up stakes entirely. Flohr's view is supported by a chemical analysis of meat storage pots during this period,[10] which showed that people changed their diets, dining more often on goats, which were hardier than cattle. Extensive knife marks on animal bones suggest they were scraped for every last bit of meat, and analysis of molecules from the animals' fat reveals that they ate plants touched by drought.

City dwellers and farm animals may have been struggling, writes Flohr, but at least some seem to have stuck around through the tribulations of a changing environment. There's evidence that people continued to live on the East Mound even as the West Mound came into fashion,[11] and social changes overlapped with the 8.2k event but weren't caused by it. Though it's not certain how much of the city's demise we can attribute to climate change, all sides in the debate agree that there was an obvious cultural shift at Çatalhöyük in the later part of the city's life. Artistic expression changed, as well as architecture, food sources, and population size. People moved from one mound to the other, or left the region altogether. They traveled a lot more between cities and villages. Slowly, a divide opened between haves and have-nots. It's possible that this may have been what finally set people on the road away from their city.

The hierarchy problem

One of the most peculiar features of Çatalhöyük's city grid is how similar the houses are. Strolling through any modern city, we expect to see houses of every shape and size, as well as apartment buildings that are warrens of tiny studios, soaring penthouses, or basement rooms whose dirty windows poke out just above the sidewalk. There are also glossy corporate towers, massive churches, imposing government buildings, and thousands of shops in every configuration. Today's cities are places where we can see social and economic inequality built into the landscape. But at Çatalhöyük, for hundreds of years, everybody lived in houses of roughly the same shape and size. Like Dido, everyone had a main room with a hearth and bed platforms, flanked by smaller rooms used mostly for storing food. Some history houses were more elaborately decorated than others, with impressive stacks of plastered bulls' heads, multiple skulls in the floor, and evocative paintings on the walls depicting hunts and celebrations. But even these fancier places were no bigger than their neighbors. More importantly, there were no buildings that did not function as dwellings. It appears that there were no purpose-built temples, nor markets. Every room, no matter how elaborate, was fitted with a hearth and a bed platform.

As Hodder puts it, there's a rigid equality to the urban design in the city. He called it "aggressively egalitarian," and suggested there may have been a taboo against keeping too much private property. There were no kings or big bosses. People in Çatalhöyük may have turned to a group of wise elders for guidance, or appointed local leaders, but those leaders did not make an ostentatious show of their authority. This is one of the many reasons why archaeologists say Çatalhöyük resembles a mega-village rather than a city. Like a village, it is a collection of houses that are roughly equal in size, with no obvious center of power. Maybe, as Tringham suggested, it looked like that because it was simply several villages planted next to each other.

That changed around 6000 BCE. Houses built during the West Mound's peak occupation were much bigger than Dido's in the East.

Single-room hearth dwellings gave way to two-story houses with many large rooms and walled courtyards. People lived in lower-density communities but built much larger areas for food storage. We no longer find evidence of people burying their dead in the floor, and there are no bones or bull skulls plastered into the walls. Pottery becomes more elaborate and highly decorated, as if people liked the idea of putting out fancy dishes for company. At the same time, we see more household items that come from distant areas. Either they're made with materials that come from far away, or they're made by people in other settlements.

It would appear that the people of the West Mound were still house-proud, but their art and symbolism was no longer connected to the structure of their homes. Instead, it was liberated from the walls and could be traded back and forth the way skulls once were. People had just as much stuff, but it wasn't all locally produced.

This shift toward larger houses and trade on the West Mound may have been a sign that social hierarchy was emerging. There were people who had two-story houses and a lot of storage, while others still had only the single-room dwellings that were once the norm on the East Mound. Notre Dame archaeologist Ian Kuijt believes that this architectural shift reveals a conflict that had long been brewing in the city.[12] People in places like Çatalhöyük, he explains, had inherited their ideas about community and spirituality partly from their nomadic forebears. Because nomadic life requires everyone in the group to share resources to survive, these groups developed customs and rituals that reinforced a very flat social structure. If anyone started hoarding resources too much, that would be bad for the entire group, so people would strongly discourage each other from ostentatious displays of social differences. This might be one reason why houses at Çatalhöyük were so outwardly similar, even though people clearly had very different amounts of food and symbolic objects in the private realm behind closed doors.

Social pressure to be egalitarian can work well in a small community where the lives of your neighbors are bound to yours. But once you have thousands of people living together, equality is harder to maintain. City dwellers might want representatives or proto-politicians to speak up for their interests, or a trade guild leader who can understand the special needs of, say, obsidian toolmakers. It's hard to make personal connections with everyone in a city full of strangers. People living in Çatalhöyük were torn between two sets of customs: the older communitarian one, in which difference and hierarchy are discouraged, and the newer, urban one, in which both are unavoidable.

Kuijt believes major conflicts would have emerged when traditional egalitarianism started to feel like rigid conformity. When tensions ran too high, people might have started to walk away from the East Mound, whose urban plan was purpose-built to promote the idea that nobody should be significantly different from her neighbors. West Mounders built homes spaced farther apart, with a wide range of floor plans, suggesting a society where people were publicly proclaiming their individuality.

Still, architectural reform wasn't enough to keep people there. Roughly three centuries after the first signs of habitation on the West Mound, almost nobody was living there or on the East Mound anymore. And by 5500 BCE, Çatalhöyük was entirely empty. Kuijt ascribes the city's demise to a broader "failure of the Neolithic experiment," and argues that it was part of a more widespread abandonment of Neolithic mega-villages throughout the Levant during the 5000s BCE. It may have been, as Hebrew University archaeologist Yosef Garfinkel put it, "a failure of the public sphere." People couldn't agree on new ways of organizing their society, and this eroded their attachment to a place that increasingly represented a dying tradition. And yet, the city did last for over 1,000 years. It may ultimately have been what Kuijt dubs "the Neolithic dead end," but the story is more complicated than that.

While Hodder and Kuijt argue that the city's early layout suggested egalitarianism, Rosemary Joyce disagrees. She's not convinced that the city's populace had ever enjoyed a "flat" social structure. I visited Joyce in her office at UC Berkeley to find out more.

Clearing a pile of books away from a chair to make room for me, Joyce wasted no time in getting to the point of questioning everything I had learned. She's profoundly skeptical about the idea that houses at Çatalhöyük all looked basically the same. She thinks that Tringham's excavations of Dido's house make it obvious that there was never one kind of "ideal" Çatalhöyük house. I recalled my conversation with Tringham when we first met, and how she traced the outline of the rooms in Dido's house on an excavation drawing. Though Dido lived during the height of what Hodder would call the city's flat social structure phase, there was strong evidence that her family had built additional rooms for more than storage. There were two small rooms off the hearth area that may have been bedrooms or workrooms. Tringham described how the doorway between Dido's main hearth room and those side rooms was walled off at some point late in the house's life, as if their inhabitants had moved on or died.

At the time, I had focused on all the ways Dido's house resembled other ones in the city. But Joyce highlighted its ever-changing number of rooms. Perhaps, said Joyce, this kind of variability was the norm at Çatalhöyük, and we've been so enamored of the "egalitarianism" hypothesis that we missed evidence of variability right in front of our noses. I had to admit she might be right. I thought about how journalists 50 years ago had gotten so excited about Mellaart's idea of a goddess-worshipping matriarchy. Maybe my fascination with an egalitarian society was blinding me to evidence of social classes. Plus,

Joyce continued, house size isn't the only way to measure hierarchy. Çatalhöyük's elaborately decorated history houses could also be a sign that there was inequality between households. "Inequality emerges from differences, even modest ones. Some houses are full of symbolic media and others aren't. To say this isn't inequality is strange." She paused and shrugged. "I'm sorry, but these houses are not the houses of equal people."

Joyce pointed out that we fundamentally don't know how people at Çatalhöyük would have understood social hierarchy. Maybe it didn't have much to do with how many bins of grain a person had, or how many plastered bulls' heads. Maybe there were shamans who wore special body paint or perishable clothing that didn't survive the millennia. Though everyone in the city might have recognized one particular shaman as a leader, we wouldn't necessarily see evidence for his status in the archaeological record. Social status doesn't always translate into material wealth, Joyce said. It can also mean access to secrets, or special places, or exclusive gatherings. And that's something we wouldn't see in the remains of a house or skeleton. "Hierarchy and access can be measured in things that don't show up in people's bodies," she mused. "Sometimes you get a hierarchical ruler without architectural signs." Joyce's perspective adds nuance to Çatalhöyük's unresolvable conflicts. One battle was brewing between traditionalists who wanted a flat social structure and those who didn't; another simmered between the emerging elites and lower classes.

Late in the city's life, Hodder said, these tensions were exacerbated by another social shift: the city's residents were becoming more mobile, traveling great distances to other cities or to quarries where they gathered raw materials. They were starting to realize that other options existed beyond the city's walls. The decorated clay stamps that people in the Neolithic commonly carried—probably as a way to display personal identity—weren't their only form of portable crafts. People at Çatalhöyük traded with neighboring communities, some over 100 kilometers away, swapping jewelry, baskets, pottery, shells, and raw materials like obsidian and chert to make blade tools. These trade networks suggest that domestication always involved social connections with remote communities, but movement between these communities became more common as the city neared its end. People's identities were less entangled with a specific built environment, and more entangled with their trade goods. As they wandered between settlements and saw a plethora of objects from far away, Çatalhöyük might have seemed less special. The place was losing its allure.

The Death Pit

V. Gordon Childe, the anthropologist who invented the term Neolithic Revolution, also created a definition of cities in 1950 that continues to influence archaeology today. To be a city, he argued, a settlement must have a population living at high density, monumental architecture, symbolic art, specialization, money and taxation, writing, long-distance trade, surplus goods, and complex social hierarchy. By this definition, Çatalhöyük is a proto-city at best. It had no money, writing, nor monuments—and probably only a simple hierarchy. Hodder says he's comfortable calling it a town, but not a city. "Çatal doesn't fit the classic city definition, if by that you mean specialization of production," he told me. "There are no zones for special activities. You don't have different parts of the city doing different things. Everything happened inside the house, from rituals to economic production."

Still, there are good reasons to think of Çatalhöyük as a city. As UCLA anthropologist Monica Smith writes, Childe's framework is

intended to define "the most complex form of aggregated populations"[13] in a relative way. We might say that Çatalhöyük was the city
of its time, more complex than any of the surrounding settlements.
Smith adds that archaeologists today also believe that it's possible to
have a city without a rigid hierarchical structure—instead, all that's
required is "highly visible labor investment, and a sustainable social
network afterward."

Smith's point here might also help us unravel the mystery of why
people left Çatalhöyük before it became a city more like Uruk, which
rose in the Levant thousands of years later, complete with writing,
taxation, money, and giant ziggurats. Essentially, the labor investment required to maintain the city and its social network was no longer worth it. This forms the essence of historian Joseph A. Tainter's
grand theory in his influential book *The Collapse of Complex Societies*.[14]
He argues that most societies lose their cohesion when people get
"declining marginal returns" on their investment in a city's physical
and social infrastructure. Rosemary Joyce put it another way: "When
you live in a city, the walls of other people's houses will fall on you.
Stuff accumulates in the street that will affect you. You take on a lot
of extra work. Çatalhöyük was an attractive investment for many
generations, and then it stopped being enough of an attraction to
make up for it going downhill." The people who lived at Çatalhöyük
toward the end would definitely have dealt with a lot of collapsing
structures, with fewer helping hands to clean things up. Walking
away was hard, but it was easier than resolving the problems that were
ripping Çatalhöyük apart.

One of the most influential post-Childe definitions of the city
comes from historian William Cronon, who argued in *Nature's
Metropolis*[15] that a city is defined partly by the rural and agricultural
regions that support it. Though Cronon was talking about Chicago,

an industrial metropolis, his idea is crucial for understanding Çatal-höyük's status as a city. In essence, Cronon made the case that agri-cultural complexity is a key part of urbanism. We know that people at Çatalhöyük grew a wide range of crops and reared animals, and processing farm products would have taken a lot of time. Troubles created by the 8.2k climate change event, followed by course shifts in local rivers, would have been major sources of strain. There's ample evidence that the city wasn't abandoned entirely because farms were failing, but food insecurity would have sent some families packing.

Though Çatalhöyük occupied a gray area between city and proto-city, its abandonment fits a pattern that emerges repeatedly in urban history. Climate change made farming tougher, and the city's festering social and cultural wounds drove neighbors away from each other. Each abandoned house created more work for the people who remained behind, trying desperately to prevent those rooftop sidewalks from collapsing under them. Over time, individual acts of abandon-ment added up to a mass action. But this process would take centuries. For people living through this late period in the city's history, it might not have been obvious that the place would one day stand empty.

Many people left Çatalhöyük to return to village life in settle-ments scattered across the Konya Plain, while others were drawn to cities like the one they left behind. British archaeologist Stuart Campbell tells a terrifying story about one of these places. He was excavating at a site called Domuztepe, roughly 130 kilometers east of Çatalhöyük, when he and his colleagues uncovered the remains of a ritual so grisly that they named it the "Death Pit." Founded just as Çatalhöyük was nearing the end of its life, Domuztepe boasted thousands of residents and lasted for centuries. It was one place where mass society flourished after most of Çatalhöyük's residents had abandoned their homes, and we can see evidence of some continuity with West Mound culture. Houses were large, often decorated with red ochre, and didn't contain any skeletons in the floors as they had on the earlier East Mound.

In fact, Campbell and the excavation team found almost no skeletons at all until they uncovered the Death Pit, which held the remains of about 40 people. Many of the bones came from older skeletons, likely ancestors, shattered and mixed into a thick cement-like clay that was layered on top of the remains of several whole cows and other animals from a feast. After the feast, people had molded the bone-studded clay into a hollow mound and stuffed it with more human remains, this time from the recently dead, including several skulls that appeared to have been smashed on one side. Campbell and his colleagues speculated that the skulls were crushed in this way so their brains could be scooped out directly after death.

Domuztepe's people dug the Death Pit next to a very special place in their city: a 75-meter-long "Red Terrace" that cut through the center of the settlement. The Red Terrace was probably over a meter high, perhaps creating an elevated walkway or ceremonial boundary, and Domuztepe's people had built it up over hundreds of years using layers of imported red clay interleaved with white plaster. It would have stood out in the city's landscape, a startling red-and-white wall dividing one side of town from another. To dig the Death Pit, residents at Domuztepe had to scoop out part of the Red Terrace and dig beneath it. After whatever activities led to all those dead bodies, the city's residents lit a bonfire so enormous that it left a thick ashy layer. Campbell and his colleagues speculate that the fire would have been visible from far away. This monstrous flame, roaring over a structure made from human bodies and clay, was surely a symbol of the city's might. Campbell described it as an example of how people entangled their identities with a location, echoing practices we see in Çatalhöyük's houses.

The difference is that the Death Pit was created as part of a large-scale public ritual. This isn't a set of ancestor bones buried in the floor of a home, for the benefit of a domestic group; it's a large collection of bodies, some freshly killed for the occasion, buried next to a wall that spans almost the whole city. We could think of Domuztepe's urban grid as an example of late Neolithic people trying to move beyond the world of Çatalhöyük. Here, boundary walls didn't merely carve out a private sphere. Rather, they were fetishistically molded into a central public monument, suggesting a fascination with difference and hierarchy. And the Death Pit ceremony itself must have been led by a person or group of people with enough authority that they could gather the city together for a multi-cow feast, ancestor ritual, and sacrifice. The hierarchical structures we see starting to emerge at Çatalhöyük appear to have been in full flower at Domuztepe. As Campbell told the story, I kept thinking about the people at Çatalhöyük who chafed against the city's flat social structure. Perhaps some of them abandoned Çatalhöyük to join Domuztepe.

In a paper about the Death Pit,[16] Campbell points out that the bone-and-fire ceremony wasn't some gory act of violence. Instead, it was a transformative ritual, joining people to the land. The Death Pit was a place where human remains were treated like clay. The bones of humans became the bones of the city. In the modern world, Campbell explains, we make a rigid distinction between the living and the dead, the animate and the inanimate. But our categories "may not reflect past beliefs," Campbell writes. Maybe the Death Pit was also an act of life-giving, in which human blood and bone symbolically revitalized the city.

In the Neolithic imagination, houses and cities might be equivalent to people and societies. The city is alive. It is a tool, an ancestor, a cosmology, and a history. When we leave it behind, we leave a part of ourselves behind, too. But when we walk the streets of the next place, we find ourselves again, for better and for worse.

PART TWO

Pompeii

THE STREET

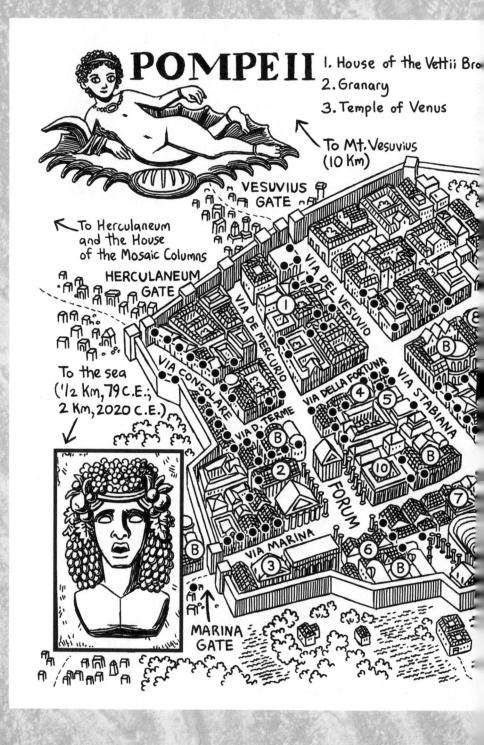

POMPEII

1. House of the Vettii Bro
2. Granary
3. Temple of Venus

To Mt. Vesuvius (10 Km)

VESUVIUS GATE

To Herculaneum and the House of the Mosaic Columns

HERCULANEUM GATE

To the sea ('1/2 Km, 79 C.E.; 2 Km, 2020 C.E.)

VIA DEL VESUVIO

VIA DE MERCURIO

VIA CONSOLARE

VIA D. TERME

VIA DELLA FORTUNA

VIA STABIANA

FORUM

VIA MARINA

MARINA GATE

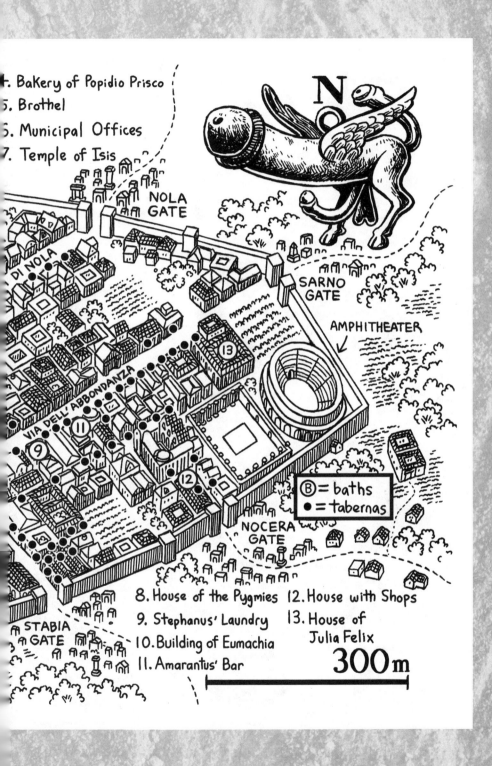

4. Bakery of Popidio Prisco
5. Brothel
6. Municipal Offices
7. Temple of Isis

N

NOLA GATE

DI NOLA

SARNO GATE

AMPHITHEATER

VIA DELL' ABBONDANZA

⑬

⑪

⑨

⑫

Ⓑ = baths
• = tabernas

NOCERA GATE

STABIA GATE

8. House of the Pygmies
9. Stephanus' Laundry
10. Building of Eumachia
11. Amarantus' Bar
12. House with Shops
13. House of Julia Felix

300 m

CHAPTER 4

Riot on the Via dell'Abbondanza

About 5,000 years after Çatalhöyük's population trickled away from the West Mound, a city of roughly the same population was buried under six meters of broiling hot volcanic ash. Unlike the people of Çatalhöyük, who gradually drifted away by choice, the 12,000 residents of Pompeii suffered an abrupt loss. In 79 CE they watched their city obliterated by a terrifying, violent volcanic eruption that must have haunted them for the rest of their lives. Earthquakes moved the shoreline a kilometer away from Pompeii's city walls, and Mount Vesuvius spewed a thick layer of fiery ash that turned fertile farms into sterile wastelands. After the disaster, refugees from Pompeii fled to the nearby coastal cities of Cumae, Neapolis (Naples), and Puteoli (Pozzeoli). Only one first-person account of the catastrophe was ever recorded.

It wasn't until the 1700s that engineers working for Charles VII, King of Naples, began to excavate the city systematically. The place was a revelation because it had been preserved intact beneath the hardened ash. Other Roman ruins had fallen into eroded piles of marble, or were buried beneath modern cities. But at Pompeii, everything was preserved, from sumptuous temple offerings to price lists for takeout. Early explorers carefully chronicled what they found but focused most of their energies on plundering gold, jewels, and priceless mosaics. Today, however, archaeologists come to Pompeii to glimpse everyday life at the height of the Roman Empire. The city was frozen in time, or perhaps cooked, with all the quirky cultural ephemera that are usually erased in continuously occupied cities like Rome or Istanbul.

The house was arguably the center of life at Çatalhöyük, but at Pompeii the street was where everything happened. In the shops, baths, and tabernas (pubs), people lived and worked, made plans, and met new friends. Romans invented a new kind of public life in their streets, codified by law and enforced through social norms. People of all classes and backgrounds mingled on sidewalks made from cement and compacted dirt. Old villas belonging to the ultrarich sprawled near a business association for freed slaves; well-heeled tourists from three continents rubbed shoulders with townie bartenders in the tabernas. Rich landladies glanced sidelong at sex workers calling to men from the rooms where they plied their trade. Nothing embodied everyday public life in Pompeii more than the city's street scenes and related amusements.

Pompeii was snuffed out at a pivotal time in Roman history, when the old social hierarchies of the Republic had crumbled away, and radical new ideas were springing up in their place. Ordinary people could challenge the supremacy of Rome's aristocratic elites and win. Women became entrepreneurs and public benefactors, while former slaves got rich. There was social mobility. When the eruption darkened the skies overhead with ash, Pompeiians were in the middle of a slow social revolution. On its streets, smeared with smutty graffiti, full of bars and bathhouses and brothels, we can see the footprint that these changes left behind.

Isis and the Pygmies

Pompeii's history begins in the fourth century BCE.[1] A bustling port city on the Bay of Naples, it was ruled by the Samnites, Rome's uneasy allies. Its residents spoke Oscan and built temples to Samnite gods, farming the fertile volcanic soils on the slopes of nearby Mount Vesuvius. They fished in the bay and traded with cities across the Mediterranean. Economically rich and strategically located at the nexus of the sea and a large inland river network, Pompeii was an obvious target for Roman conquest. But for at least two centuries, Rome was content to treat Pompeii as an ally, as long as the town provided soldiers for its wars. Then, in 91 BCE, Pompeii and a few other southern Italian towns sparked the so-called Social War with Rome, partly in an attempt to gain more rights after centuries of serving as de facto client states.[2] After a bitter struggle, a Roman army led by Lucius Cornelius Sulla crushed the Samnite resistance in 80 BCE. Pompeii became a fully Roman city, and Sulla forcibly settled thousands of retired Roman soldiers there. The new Romanized population converted Samnite temples to Roman ones,[3] and Pompeii's official language became Latin.

This colonial history set the tone for Pompeii's polyglot culture. Though technically Roman, Pompeii still had a thriving Samnite community who openly worshipped Oscan deities like Mefitis, a multifaceted goddess often compared to Venus. People kept scribbling graffiti in Oscan on Pompeii's walls right up until the day Vesuvius erupted. Immigrant cultures also thrived throughout the city, and one of the strongest non-Roman influences came from North African empires.

I arrived in Pompeii at the height of summer, stepping off the train from Rome along with a group of bored-looking schoolchildren who were clearly there for the same reason I was. The modern city of Pompei—spelled with one "i" instead of the ancient city's two—caters largely to tourists interested in the ruins. Most visitors pause by the stalls hawking tinfoil Roman helmets and gelato, then make a beeline for the sumptuous villas overlooking the sea to the west. But I began my visit by wandering through its quieter southern neighborhoods, looking for the impact of North Africa. I was very close to the Forum, a formal downtown area where Pompeiians built administration offices and at least a dozen temples. Among these monumental buildings, I found the crumbling glory of the Temple of Isis, an Egyptian goddess, whose dais and once-colorful columns are now a uniform gray stone. All that remains of this luxurious retreat is a walled enclosure, whose generous dimensions hint at all the money poured into its shrines and dwellings for its priests. Its lavishly painted frescoes depicted Isis worshippers' life along the Nile, and are now in the permanent collection at the Naples Archaeological Museum. In the first century, Isis worship was all the rage in Pompeii, and rich Roman women were especially keen on the imported African goddess.

Around the corner from the Temple of Isis is a street called Via Stabiana, a major artery that leads down a gentle slope to the Stabian Gate, one of the ancient city's main entrances. Millennia ago, this street would have been flanked by two theaters, dozens of bars, and some villas. On holidays devoted to Isis, Via Stabiana would have been packed with costumed revelers led by the women who ran the temple. But today the gate was closed for an excavation, and the further I wandered, the less I could hear the rowdy tourists visiting the city's more famous attractions. I sat on a curbstone, looking at an archway from the Stabian Gate. Behind me were the city's ubiquitous brown brick walls crumbling into plots of dried weeds and hardy wildflowers. I imagined the place thick with pedestrians, mule-drawn carts,

RIOT ON THE VIA DELL'ABBONDANZA

and merchants hawking their wares from storefronts built into the bottom levels of three-story homes above me. But now the street felt truly abandoned, devoid of life and context.

And then, as if by magic, the eminent University of Cambridge archaeologist Andrew Wallace-Hadrill appeared. A dapper figure in a linen suit with a sweep of white hair combed back from his forehead, he emerged from a small residential alley choked with waist-high weeds. It was a fitting entrance: Wallace-Hadrill is famous among archaeologists for pioneering a new approach to exploring these ruins, focusing on domestic life in houses rather than elite political maneuvers in the Forum.[4] Apparently he was in town for a conference, and had decided to check on some new excavations. Knowing his interest in Roman houses, I asked him how Africa had found its way into so many frescoes here. Though typically these paintings showcase Roman and Greek myths, Pompeii homes devote a lot of wall space to African scenes. Some of these images are worshipful, like the ones associated with Isis; others are the ancient Roman equivalent of racist Sambo caricatures, putting Africans in satirical or humiliating poses.

I was curious about a painting I'd seen at the museum in Naples, from the so-called House of the Physician. Curators described it as pygmies reenacting a classic scene from the Old Testament where King Solomon resolves a dispute over the maternity of a baby by threatening to cut it in half and give each woman an equal portion. The woman who responds by volunteering to give up the baby is revealed as the true mother. In the fresco, we see the scene with cartoonishly rendered African pygmies in all the roles. Solomon, wearing a gladiator's helmet that dwarfs his head, holds a meat cleaver over a wriggling baby. Two women watch, the darker one with a venal grin, and the paler one looking mournfully away. It looked to me like racist humor of the ancient world.

Wallace-Hadrill agreed with my interpretation, but laughed out loud at the idea that it might be a scene from the Bible. "That's what they like to call it, but it's not described as the judgment of Solomon anywhere," he explained. "It's likely a myth about an Egyptian king, but we call it 'Solomon's Judgment' because we know the Bible." He said it was common for Romans to represent Egyptian culture using pygmies in frescoes. Some of these frescoes are basically dirty jokes: one shows pygmies floating in a penis-shaped boat propelled by a river of sperm. Others convey respect, conjuring gorgeous scenes of the Nile with realistic African figures at work or performing rituals. As the Temple to Isis attests, Egyptian gods were revered in this city. This mix of worshipful and hateful images across Pompeii speaks to an acute cultural awareness of Egyptian political power. Some embraced it, and others belittled it. But nobody could ignore it.

Wallace-Hadrill added that Africa also found its way into Pompeii from the Punic empire that was centered in the regions that are today northern Tunisia and Algeria. Carthage, a Punic city, was a major trade center and had often challenged Rome for control over the region during the war-torn Republican era. We know Pompeiians did a lot of business with the Punic world because, as Wallace-Hadrill put it, "Pompeii uses coins from Ebusus [modern-day Ebiza] in enormous quantities." The island of Ebusus was a Punic territory positioned strategically between what are now Algeria and Spain, both of which fell into the clutches of the Roman Empire. Garum, a fermented fish sauce that was a Pompeii delicacy, also had its origins in the Punic world. And even Pompeii's architecture was shaped by Punic styles. A popular method of bricklaying, in which large blocks are arranged into T-shaped patterns between smaller, thin bricks, was borrowed directly from the Punic world. Indeed, this kind of brick arrangement was referred to among Romans as "Africanum," so people clearly knew its origins.

After Wallace-Hadrill said good-bye, I retraced my steps back up Via Stabiana to the Temple of Isis, looking for Africanum in the walls I passed. Suddenly it was obvious that I wasn't in some pure, distilled version of ancient Rome, preserved for thousands of years. I was in the ruins of a diverse urban community, whose population came from many places, and fused the traditions of North Africa and Rome into something that was uniquely Pompeii. And just as all New Yorkers are not the same, neither were all Pompeiians.

The business that Julia Felix built

The names of nearly all Pompeii's residents are unknown, and excavators refer to buildings as "House of the Tragic Poet" or "House of the Surgeon," in reference to art or other items found inside. Likewise, the city's streets are known by modern names given to them by many different explorers over the years. One of the few buildings whose owner's name survives is the House of Julia Felix. A sprawling property spanning an entire block on the far northeastern side of the city, it's all the way across town from the Temple of Isis. Painted on its facade the day the volcano erupted was an advertisement for shops and apartments to rent within:

> To let, in the estate of Julia Felix, daughter of Spurius: elegant baths for respectable people, shops with upper rooms, and apartments. From 13 August next to 13 August of the sixth year, for five continuous years. The lease will expire at the end of the five years.[5]

This is the only written record we have of Julia Felix. She must have been very wealthy to own an enormous estate off Via dell'Abbondanza, a major thoroughfare that cuts all the way through Pompeii from her house on the east side to the theater and temple districts near Via Stabiana. We also know that Julia's estate changed a lot in the years leading up to 79 CE, expanding to merge with the estate next door and swallowing up an alley that had once separated the two residences. The property seems to have become more commercial, too, adding a large bathhouse and a dozen tabernas.

Though it may at one time have been a private villa—the Roman equivalent of a mansion—for Julia or a previous owner, the property was gradually modified to be more like a luxury club and spa. Bathhouses in ancient Rome weren't for getting clean, though occasionally one might accidentally emerge slightly less dirty. Baths were essentially social clubs where people discussed business and news while taking a hot soak. At Julia's bathhouse, they could read scrolls of scandalous new poetry next to a shady garden fountain, or take an afternoon nap on a couch nearby. And they could have a meal or two at one of the tabernas on her property.

Julia's bathhouse faced the lively Via dell'Abbondanza, and was probably frequented by locals as well as visitors. As her rental notice hints, it was a great place to set up shop. The street was packed with foot traffic from tourists who came in through the nearby Sarno Gate to watch gladiators practice in a large, open-air gymnasium called the Paelestra, and to see games and other entertainment at the enormous amphitheater down the road from Julia's property. The area was so notorious for rowdy behavior that it was declared a public menace by Emperor Nero in 59 CE, when a deadly riot broke out between the home team gladiator fans from Pompeii and visiting fans from the neighboring colony of Nocera. The carnage was so dramatic that Nero banned gladiator games for ten years at Pompeii.

To follow the footsteps of the rioters from 59, I entered the city from the southeast, right next to the amphitheater. The place is still used for concerts, and I had just missed my chance to see King Crimson perform where Nucerians and Pompeiians once slaughtered each other over a gladiatorial game. Walking north, I passed between the amphitheater and the columns ringing the gladiators' practice field, and then ducked into an alley edged by a vineyard that would not have been out of place during the city's heyday.

Imagining the rioters close behind me, I took a left onto the Via dell'Abbondanza and found myself on the doorstep of Julia Felix's property. Broad stairs led right from the street to her door, allowing visitors to bypass the sidewalk and proceed straight into her gardens and bathhouse. I peered inside through a gate. Though the marble columns and landscaping were gone, along with her "for rent" sign, the stateliness of the place was still palpable. It's an enormous L-shaped property, occupying an entire block of Via dell'Abbondanza and turning the corner to continue for another block on Vicolo di Giulia Felice, or Julia Felix's Alley. Through the doorway I could see the atrium where she would have received visitors, and beyond it, a landscaped garden. To my left was a bar with marble-topped counters and a bathhouse. Around the corner were more bars and private rooms to let. Julia's property is called an insula, meaning an entire square block, and over half of it was given over to a lush orchard and garden for guests.

Via dell'Abbondanza is narrow enough that I could imagine a riot would have engulfed anyone walking by, and I wondered what Julia's renters were doing while it was happening. Perhaps looking out from the garden? Gulping wine and joining in? If Julia had rooms here, she would likely have been watching from upstairs, in the less public parts of the house where people typically lived. She might have seen sports fans beating each other, or looting her tabernas.

The riots of 59 were simply the most extreme version of the kind of drunken revelry that often filled the streets near Julia's house. Many of her neighbors responded by bricking up entrances to their homes on the Via dell'Abbondanza. In 79, however, Julia had recently opened up several new entrances, making room for more tabernas to service passersby as well as bathhouse regulars. Her richly appointed complex must have been the perfect getaway for tired pedestrians with some coin to spend.

It was also the first structure in the city to be uncovered when excavations began in the 18th century. Wesleyan University archaeologist Christopher Parslow has been researching Julia's property for nearly four decades, and he said the building first came to light nearly 300 years ago when a farmer discovered the tops of marble columns sticking up out of his field. King Charles VII, a Bourbon royal from Spain who ruled the Bay of Naples at that time, had already funded excavations at two other ash-buried Roman cities, Herculaneum and Stabia. Known for his Enlightenment values, the king was fascinated by ancient history and dispatched Swiss engineer Karl Jakob Weber to examine the farmer's discovery. Digging deeper, Weber revealed a row of gorgeous marble columns—the only marble columns found in Pompeii to date—as well as the impressive building surrounding them. We know from Julia's advertising notice that the place once had extensive upper floors, but it's likely that Weber's pick-and-shovel excavation techniques ruined whatever remained of them after the eruption. Still, it was enough to convince the king that further excavations should be done. Though the House of Julia Felix was later reburied, only to be uncovered again in the 20th century, her property was what first brought the world's attention to the ruins of Pompeii.

And yet we still know very little about the woman listed as the building's owner on the sign outside. Parslow told me that for a long time, people believed she'd lived in the house until the day of the eruption. A skeleton with jewelry was found in the garden, and it was assumed to be Julia. "I don't think she was the skeleton," Parslow said wryly. "We don't even know if [the skeleton] was female." He's not convinced she lived on the premises. "Her house has a design similar to private houses, but it's much too public in terms of the interconnectedness of spaces inside. There's not much privacy in what was supposed to be a private house." If she did live in the house, "Where's her bedroom?" he asked. "There is no place to put it because there's so much traffic" from visitors. He speculates that Julia managed her substantial property from somewhere else, perhaps another estate in Pompeii. As for the House of Julia Felix? "I think it's set up for entertainment," he said. "It's for people coming in for meals and to be entertained [at the baths]."

But Julia wouldn't have wanted just anyone traipsing onto her property. "It had a level of elegance," Parslow said. It wasn't simply the marble columns; the entire property was artfully painted, and the garden boasted exquisite landscaping details like little bridges over its central fountain. He added that a public bathhouse like this would have charged a steep fee, underscoring its resemblance to an exclusive club. As her advertisement indicated, Julia was making money partly by catering to "respectable people."

Julia had to surmount many barriers to own this property. At the time she was alive, Roman law held that a woman had to manage property through a "guardian," usually her father.[6] However, it doesn't appear Julia was in that position. The rental notice mentions her father's name, but clearly stipulates that she owns the property and will negotiate with any would-be renters. One likely explanation for this comes from the so-called Julian Laws, created by Emperor Augustus in 17 BCE to govern the sexual and reproductive behavior of women. Under Julian Law, a freeborn woman like Julia could gain the right to manage her own property if she had three children. If Julia was a freedwoman, or former slave, she would have had to bear four children to achieve the same status.[7] Assuming Julia came from the moneyed classes, she likely married as a young teenager and possibly had a couple of husbands by the time she was in her 20s. Wars frequently made young brides into widows, and divorce was also widely accepted for any number of reasons. So we're left to guess that Julia probably bore three or more children, leaving her in a position to manage an inheritance from a dead husband.

The Julian Laws sound preposterous from a modern perspective, but Roman leaders took them very seriously. They would have cast a shadow over the fortunes of people like Julia. Augustus had styled himself a social reformer, attempting to curb the decadent habits of youth during the last days of the Republic. Along with incentivizing women to have as many babies as possible, the Julian Laws also meted out harsh punishment for women deemed "promiscuous." Famously, Augustus exiled his own daughter in 2 CE when she refused to stop publicly engaging in the ancient world's equivalent of free love. At the same time, an odd liberalization crept into the Roman world through these laws. To encourage marriage, Augustus allowed freeborn men to marry freedwomen for the first time, making their children legitimate. Now a woman born a slave might be freed to become the wife of a citizen, and her children would be freeborn.

We can't fully appreciate the built environment of Pompeii without understanding how it was shaped by women, and why. Though women couldn't run for office or vote, they could own property. They could become entrepreneurs or patrons of powerful cults. The outsized footprint occupied by the Temple of Isis and the property of Julia Felix are both testimonies to female power at Pompeii, and how women were reshaping the cityscape.

Nero actually did a few good things

By the time Julia Felix was running her own establishment in the city, the Julian Laws had been challenged and modified by at least three generations of young people. The restrictions on women had waxed and waned. Emperor Nero, especially, seemed disinclined to enforce rigid ideas about women's modesty. Lisa Hughes, a professor of classics at the University of Calgary, studies theaters at Pompeii during this period. When I sat down with her at a café to talk about what she'd discovered, she said something that took me by surprise.

Tucking an unruly strand of brown hair behind an ear, she gave me a sly smile. "I love Nero!" she enthused.

I was so startled that I spilled my coffee, narrowly missing the laptop I was using to take notes. "That's not something you usually hear about Nero," I laughed, as she helped mop up the mess with a wad of napkins.

Hughes shrugged, as if she gets that reaction a lot. "He was actually great for women."

Nero came from a long line of powerful female icons. His mother Agrippina the Younger became a notorious political figure after writing a popular memoir about the history of her family, including her brother Caligula, her husband Claudius, and her mother Agrippina the Elder, who was very close to the Emperor Augustus and got swept up in early Imperial political violence. As Nero took power, women were moving from the domestic sphere into the public realm. They challenged the Roman tradition that good women stayed at home weaving cloth for their husbands, fathers, and sons.

Though widely reviled as a decadent tyrant, Nero was also a populist who loved theater and music. He acted in plays, and used theatrical productions to make political arguments that previous leaders would have made in the Forum. It's not a stretch to compare Nero's techniques to those of contemporary American politicians, who use social media or television ads to spread their ideas rather than giving formal speeches. During his reign, Hughes explained, Nero poured money into theaters, and the demand for performance troupes reached a fever pitch. As a mostly unintended consequence, "under Nero, the theater is opened up and more women enter the performance space," Hughes said. Female performers became common, but women also went into the theater business as producers and patrons. Pliny the Younger, a Roman commentator who survived the eruption at Pompeii, wrote somewhat disapprovingly about a wealthy matron named Ummidia Quadratilla,[8] who owned her own mime troupe.

At Pompeii, where there were two public theaters and entertainment was a major lure for visitors, Hughes studies one of the lesser-known trends in first-century Pompeii performance: backyard theaters in private homes. There are 11 known backyard theaters in Pompeii, and Hughes speculates that wealthy people staged special performances in them during al fresco dinners, inviting select groups of friends, business connections, and political allies. Hughes sees this phenomenon as part of a larger shift in public perceptions of women's roles. It offered business opportunities to women like Ummidia Quadratilla, whose mime troupe probably performed in backyard theaters.

But theater also offered women more than economic self-sufficiency; it was a place where Romans reimagined gender roles.

Hughes told me that in the decades when Julia and Ummidia ran their businesses, there was a surge in popularity of stories about Hercules and Omphale, the Queen of Lydia. At least two houses at Pompeii boast elaborate frescoes showing a key scene from the Hercules and Omphale myth, in which Hercules gets drunk and puts on Omphale's clothes. The queen, for her part, wears Hercules' clothing or bears his weapons. In one fresco from the House of the Prince of Montenegro, she sits at the head of the table in a spot normally reserved for male hosts. "She's assuming the role of the *domus*, or male patron," Hughes mused. "It's a representation of women running the show, and running the house."

The Omphale myth reminds her of a real-life woman from Pompeii, Eumachia, who was likely the age of Julia Felix's mother. Eumachia rose from a nonaristocratic background to lead one branch of a powerful civic organization for freed slaves called the Augustales. She also became a matron of the fullers' guild, a trade group that represented garment makers who fabricated, dyed, or washed clothes. With her tremendous self-made wealth, Eumachia dedicated a large public building—known today simply as the Building of Eumachia—in a choice spot located next to the Forum in Pompeii's equivalent of a downtown area. Rome may have been officially a patriarchy, but Eumachia managed to stand in a man's place and flourish. Perhaps she thought of Omphale as a role model.

Not only did the Omphale myth challenge ideas about gender, but it also overturned ideas about ethnicity. Omphale was a foreign queen, from a region in what is today western Turkey. Hughes believes that this myth was perfectly suited to the changing values of a town like Pompeii, which was packed with immigrants from all over the Roman

Empire—some slaves, some free. "The very fact that these images were created attests to the fact that there was an audience and community that would accept them or enjoy them," Hughes said. The audiences for frescoes and backyard theater performances weren't merely public elites—they were the slaves and freed people who worked in these domestic spaces. At a time when former slaves and women were entering the public sphere in growing numbers, Hughes explained, "they're trying to establish their identities, but they aren't emulating elites." Instead, they might be taking inspiration from art and theater they saw in the domestic spaces where they were employed. "Theater is a key venue for promoting social change," Hughes said. And Pompeii was a theater-loving town.

The people in the kitchen

If you walk north from Pompeii's downtown theater district, you'll eventually pass through the city gates on a road known as the Via Consolare. Wide and well used, Via Consolare cuts through the city grid diagonally, continuing into the suburban countryside, past the enormous villas of Rome's rich and famous. This was the street that connected Pompeii to nearby Herculaneum, a smaller and more elite seaside town that was also buried in ash by the eruption of Mount Vesuvius. Some archaeologists believe the famous orator Cicero had an estate on Via Consolare about 150 years before the city was buried. The villas preserved here in 79 were just as fancy as those in Cicero's time, fit for emperors and elites. Early one morning, I walked Via Consolare, passing the crumbling remains of estates that stretched for city blocks, their once-fine atriums reduced to bare stone. Each atrium had an impluvium, or pool, at its center, now chipped and empty. Impluvia were luxurious water features intended to impress visitors, but they had a pragmatic purpose, too: each would have been filled with rainwater that fell through an open skylight called a compluvium. Nearly every Roman house had some kind of impluvium/compluvium setup, but the ones along this road were enormous, befitting their placement in majestic villas.

Tombs also stood along Via Consolare, memorials to anyone whose relatives could pay for a monument. Romans put their graves outside city walls partly for spiritual reasons, and partly so that visitors would learn about the city's most powerful families before passing through its gates. Two thousand years ago this road would have been swarming with pedestrians and carts traveling between Herculaneum and Pompeii, many gaping at the same buildings I could see around me.

But I was hunting for one particular structure, the House of the Mosaic Columns, where San Francisco State archaeologist Michael Anderson and his team were excavating. Anderson manages the Via Consolare Project,[9] a long-running investigation focused on how people used this street. I wandered back and forth, consulting my map, trying to figure out which deserted ruin was the House of the Mosaic Columns. Unlike most buildings in the park, it had no obvious markings or signage. All I could find was a gate blocking the entrance to an arched tunnel leading to an overgrown garden.

"Hello?" I called. I couldn't believe this was the place because there was no sign of anyone. But presently Anderson poked his head around the corner of the tunnel's far end and waved. After unchaining the gate, he guided me through an entrance guests would have accessed from the street 2,000 years ago. The arched walls would have been covered in elaborate frescoes, and the passageway was wide enough to admit a cart. We emerged into a walled garden so spacious that it contained an entire outdoor archaeology lab. A shade structure covered a makeshift work space with a table, computer, boxes of ceramics, and a very tidy array of hard hats. Students and other researchers passed in and out of the garden on their way to trenches they'd dug inside what was once the villa proper. Out in the garden, they catalogued ceramics and sifted through dirt in the very spot where patricians once enjoyed the intricate mosaic columns that gave the place its name. Now there were four fake columns made of bare concrete, crumbling and leaning at odd angles.

"The mosaic columns were taken to the Naples Museum and replaced with these bizarre, crappy concrete things," Anderson explained with a wry grin. He'd been working in the heat all morning, and his dark hair was held back with a bandanna. There was a dollop of sunblock on his ear, as if he'd been applying it distractedly. "I don't know why they bothered with these fake columns. They're not the right size, and they're not in the right place."

It was a lesson in the realities of Pompeii archaeology, where researchers often find themselves excavating previous reconstructions of the city as well as the city itself. Anderson said he's found display cases and fake bits of architecture from the 19th century, as well as the 1910s and the 1950s. He showed me what looked like two rusted metal cages in the grass, roughly the shape of coffins, which are all that remains of a 1950s display of Samnite graves that lie beneath the villa's Roman incarnation. "There's this one moment when this villa is a showpiece in the 1950s. We have pictures of it. They repiped the water, and a fountain was running in the garden here. There were trees, and the fake columns. Then it was very quickly forgotten." When he excavated the doorway to the villa, he found a lintel stone that had been placed in 1910. "You have to be prepared for anything," he said.

After 14 years running the Via Consolare Project, Anderson is grappling with two mysteries. What was this villa like during the city's heyday? And how did previous digs alter its appearance for the sake of 20th-century tourists? Pompeii is a meta-archaeological site, revealing ancient history right alongside the history of archaeology as a field.

Though earlier generations of archaeologists were fascinated by this villa's wealthy owners, Anderson is part of a new wave of researchers influenced by Wallace-Hadrill who are interested in the domestic lives of ordinary people. That's why he immediately took me through a door from the garden into the villa's kitchen, where slaves, and freed slaves called *liberti*, would have worked. This was a luxurious space even for such an enormous villa. Anderson was astonished to discover the kitchen had four cooking platforms, which is incredibly rare. Most villas have two at the most. And this kitchen was roomy enough for at least a dozen people to work comfortably, with ample storage rooms adjacent to the central work area. Anderson and his colleagues were especially excited when they realized it had been replumbed with lead pipes during the first century. The pipes fed a fountain installed in the corner of the kitchen, providing endless fresh water for the chefs. This alone would have been an incredibly expensive and unusual feature, and Anderson wondered aloud why the villa owners would have commissioned it. One possibility is that they hosted a lot of feasts for guests. Another possibility is that the kitchen was also being used by the merchants selling food to passersby.

We walked back out through the garden archaeology lab to get to Via Consolare, where Anderson described what the street would have looked like in 79 CE. As he talked, his fingers hovered in the air as if he were drawing an architectural diagram. The House of the Mosaic Columns stood at least three stories above the street, partly thanks to landscaping that propped it up with an artificial hill behind the garden. An open-air peristyle dominated the top floor, boasting a beautiful ocean view. This topmost level was built with a slight over-hang, providing shade for the colonnaded sidewalk below. Anderson noted that the overhang also had the effect of blotting out the street view for people inside the peristyle, so they could enjoy the ocean view without seeing the dirty scene below.

Like Julia Felix's property, this villa was built with shops facing the street on the ground floor. From Via Consolare, the House of the Mosaic Columns would have looked like a long row of retail outlets, kind of like a strip mall, with a couple of entrances between the store-fronts leading to the villa's hidden garden and kitchens. Anderson showed me where a bronze metalworker's shop once stood, alongside food stalls, bars, and more. "This is the longest single block of shops in the city other than the Via dell'Abbondanza itself," he said. "That's why I was drawn to this place." He paused, lost in thought, and I imagined the shops around us filled with life, the sidewalk hazy with smoke from the blacksmith's fire and spiked with the smell of cumin and coriander sizzling with fish in olive oil. "The villa is supported by these businesses monetarily, but also supported literally. It is its own metaphor," he mused. "I don't think that's an association that would have been lost in antiquity."

"Certainly not to the people living on the bottom, propping it up," I replied.

Anderson chuckled and nodded. "For me, this isn't about the Julius Caesars and emperors who we know too much about already—it's the people we don't know anything about. Even if we never know their names, we can manage to reconstruct a bit of their lives."

So who were the people who rented these shops from the villa's owners? Most likely they were *liberti* whose lives were joined to their former owners by Roman law. Once a slave was freed, his or her mas-ter became known as a patron, and patrons generally took on the role of adoptive fathers to their former slaves.[10] A *libertus* (or, if female, *liberta*) would usually remain with the family that had once enslaved him or her, running independent businesses or helping to manage the patron's estate. Cicero wrote that his beloved *libertus* Tiro managed all the politician's business affairs. By most estimates, Roman cities were thronging with slaves and *liberti*, partly because the patronage system made freeing slaves an economically appealing proposition. The villa's owners would lose a piece of human property, but they would gain a loyal worker whose fortunes were tied to that of the household. Roman manumission was often slavery with benefits.

King's College historian Henrik Mouritsen, author of *The Freedman in the Roman World*, estimates that the typical Roman household would have been roughly half slaves, and a quarter to a third *liberti*.[11] Anderson agrees. He speculates that the people running the shops "could all be slaves, or former slaves, or distant family relations." Probably all of the above. Family connections were key to Roman life, and a household that required four cooking surfaces was probably quite large, full of many relationship permutations. Someone born a slave at Pompeii could climb nearly to the top of the social hierarchy, if he were lucky enough to be freed and given a nice position by his patron.

Still, the growing *liberti* class was blocked from holding positions of political power. They had to content themselves with gaining status through civic organizations like the Augustales, a group created by Emperor Augustus for *liberti* who wanted to gain business and political connections. A *libertus* might hope that his children could vote, but he never would. Social hierarchy was under reconstruction, but the power gap between male elites and everyone else still smarted like an infected wound.

There were also more subtle signs of class warfare. University of Reading classics professor Annalisa Marzano has uncovered evidence of what seems like a peculiarly modern conflict between the haves and have-nots over access to the beaches on the Bay of Naples. Rich vacationers built enormous villas on the shoreline. Instead of building tabernas into their lower floors, they constructed aquaculture tanks that stretched into the sea.[12] Marzano believes the idea was to lure even more fish into their artificial ponds. Presumably, slaves and *liberti* were in charge of managing these tanks, selling fish that their patrons didn't eat at local markets in Pompeii, Herculaneum, and elsewhere along the coast. As this kind of villa became more popular, local fishers found their access to the beaches blocked. The families inside were stealing their livelihoods.

Tensions came to a head a few decades after Vesuvius had blotted out Pompeii. Legal documents show two fishers petitioned the emperor Antoninus Pius to intervene on their behalf because villa owners were preventing them from fishing off the coast of their hometowns. Pius declared that anyone could have access to the sea, with one exception: nobody would be allowed to fish near the villas. Pompeii's demise came right in the middle of an ongoing conflict between rich and poor, men and women, immigrant and Roman and indigenous. We may not know the names of individuals involved in these conflicts, but they left their marks on a city where Romans built with Africanum, female entrepreneurs funded the Augustales, and elite villas rested atop the tabernas of freed slaves.

CHAPTER 5

What We Do in Public

P ompeii was already something of a disaster magnet when Vesuvius erupted in 79 CE. Seventeen years earlier, the Bay of Naples had been hit by an earthquake that leveled large parts of the city and sent a tsunami hurtling into the nearby Roman port of Ostia. Many of Pompeii's residents left and never returned after the quake, leaving behind damaged buildings that were still empty in 79. In a sense, Pompeii's abandonment begins with that quake, which shrank the city's population and eroded its desirability as a vacation spot. Still, many people stayed after the dust had settled, eager to renovate and upgrade. Emperor Nero helped fund the restoration efforts, which are still visible in walls whose exposed brick shows extensive repairs. Remains found at the Sanctuary of Venus, a temple dedicated to the city's patron goddess, reveal that engineers in 79 were reinforcing it with thick stone walls that they believed would be earthquake-proof.[1] The Pompeii that we see today was a city under construction. Land owners were redesigning their properties to reflect a more modern Roman sensibility, oriented around trade rather than military conquest.

Put another way, Pompeii's post-quake urban landscape was geared toward shopping. *Liberti* and other nonelite groups transformed many of Pompeii's great villas and homes into mixed-use spaces, with retail infringing on what were once living quarters. It's likely that the renovation of Julia Felix's property was part of this trend, as she added more doors to face the commercial Via dell'Abbondanza. Retail-oriented transformations were taking place all over the city, and we see laundries and bakeries springing up in what were once the fancy atria and gardens of elite estates. Most of all, we see tabernas, the bars and restaurants that were Pompeii's ubiquitous retail outlets. Some tabernas were tiny, one-room affairs with a counter, and others boasted many rooms and garden seating. Their proprietors served hot food, cold takeaway, and wine of various types. I saw the remains of several at Julia's place and the House of the Mosaic Columns, but that was just the beginning.

Every major street in Pompeii is lined with tabernas, and I learned to recognize them from their characteristic L-shaped marble-topped counters. These counters always came with built-in ceramic storage containers, roughly 60 centimeters deep, their wide, round mouths level with the countertop. Many were probably covered with wooden lids that are long gone. Peering inside one, I could see the smooth inner walls of what was once a display case for dry goods like grain or nuts. Today they look stark and minimalist, almost like an ellipse punctuating the countertop. But frescoes from the time show taberna counters piled with goods, while herbs, fruit, and meat hang from the ceiling over the counter. Amphorae, elongated clay carafes for wine, olive oil, and other liquid goods, lean on their pronged feet against the walls. Tabernas would have been utterly astonishing to the Neolithic people of Çatalhöyük, who produced every ingredient in their meals, from pots and hearths, to spices and proteins. At Pompeii, a person simply had to walk outside to get meal service—or to visit specialists in cookware, oils, meats, and vegetables to make a meal at home. There were even shops that sold hot water, in case you didn't want to boil your own.

Taberna hopping

To talk about Pompeii's tabernas, I sat down for beers with longtime collaborators Eric Poehler, from the University of Massachusetts at Amherst, and Steven Ellis, from the University of Cincinnati. Ellis is the author of *The Roman Retail Revolution*,[2] a book chronicling the rise of small businesses in the Roman world. He has studied tabernas across the empire, from North Africa to the Middle East, and he told me that Pompeii had over 160 tabernas. "It's an extraordinary number," he said, adding that it's probably a low estimate because parts of Pompeii are still buried. Poehler followed up with some back-of-the-napkin calculations: "If you've got 160 bars for a population of 12,000 people, then about a tenth of them had to be eating out to sustain the bars." Who were those people? Wealthy residents would have had a whole fleet of slaves preparing meals in their well-appointed villa kitchens. The city's humbler dwellings did not have kitchens, so it might seem at first glance that tabernas were for the poor. But that doesn't fit the evidence, either. Even in a tiny upstairs apartment with no running water, people of little means could cook using a pan over a small brazier, like using a hot plate.

Ellis believes that these tabernas were run and frequented by a group he calls "middlers," people who aren't rich or poor. They ate outdoors on the same blocks where they ran small shops that sold everything from onions and fish sauce to textiles and perfume. Most of the middlers, Ellis said, "have money to spend on food" and other small luxuries. They aren't precisely what we would call middle class because that's a term associated with modern societies. Indeed, some are quite wealthy, while others are recently freed *liberti* just trying to scrape by. But they are all in the vast economic middle between Roman elites and slaves. What truly sets these middlers apart, though, is that they earn money by working in business or a trade. This kind of work was taboo for elites, though of course many of Rome's richest made their money from shops and farms staffed by their *liberti* and slaves. As Anderson pointed out at the House of the Mosaic Columns, the villa was literally and figuratively supported by shops built into its bottom floor.

Newer Pompeii houses reflect a city whose wealthiest members were also working for a living. The building known as the Fullery of Stephanus appears to have been rebuilt after the quake, converting an elite house with an atrium into a "fullery," or wool processing workshop. Stephanus' name was discovered in an election-related sign painted near the fullery's entrance on Via dell'Abbondanza. (Though we can't be sure that Stephanus was the name of the baker who lived there, we'll assume it is for simplicity's sake.) It seems that Stephanus dramatically retrofitted the patrician house, converting the tiled atrium into several large, utilitarian rooms full of tubs and tools for treating wool. But he also left considerable room for domestic life. Leiden University archaeologist Miko Flohr, author of a book about ancient fulleries, has explored Stephanus' property and found jewelry, cosmetics, cookware, and other signs that people were living there as well as working. Instead of segregating the "work" part of the house into a strip mall facing the street, Stephanus integrated it into his home, where his slaves and *liberti* no doubt lived with his family.

"Essentially," writes Flohr, "it was just a house in which people lived, slept, ate, and worked, and which they probably considered home."[3]

Further along the Via dell'Abbondanza we find a similar late reconstruction, known as the House of the Chaste Lovers, where a patrician villa was rebuilt so that its entrance led directly into a big, airy bakery. Ovens and millstones dominated the spaces that once would have shaded hushed conversations between aristocratic people of leisure. The baker cared a lot more about keeping the mules who ran his millstone happy than he did about luxurious living. He built a stable next to the former residents' triclinium, a dining room where slaves served their patrons supper as they reclined on formal dining beds. Like Stephanus' fullery, this bakery integrated work spaces with domestic ones. It was connected to a residential house, but Poehler said the owners prioritized rebuilding the bakery, finishing those parts of the construction first. The people who owned this bakery may have been rich, but they were not part of Rome's patrician class. They worked for a living, and labor was literally a part of their household.

Stephanus and the anonymous bakers who lived in the House of the Chaste Lovers were probably the main audience for Pompeii's dizzying array of bars and restaurants. Ellis calls this period Pompeii's "retail revolution." Civil strife in the empire had petered out, and Romans were enjoying a rare period of peace. "This is the beginning of halcyonic Rome, and the volume of trade is spiking," Ellis said. "We're moving from crafts being done at the individual level, to individuals participating in a craft industry at scale." At Pompeii, this means that people aren't just buying and selling to each other. They're part of a vast economic network that stretches all across the empire and into Africa, Asia, and the Middle East. On Via Stabiana, Ellis has identified tabernas that reflected this new cosmopolitan reality. Looking at evidence from storage bins, cesspits, and menus, researchers found that one taberna had a very simple menu of locally grown fruits, grains, and vegetables with some cheese and sausages. The other, two doors away, had a much wider variety of foods. "[There were] cumin, peppercorn, and caraway coming from India," he said. "The food was flavored with spices that are foreign." Middlers going to a Pompeii taberna could choose to dine on imported delicacies or local comfort food options. Delicacies once available only to elites were now part of everyday life for more people than ever before. Even people born into slavery might eventually have their own shops and dine like patricians. Interestingly, data from the modern world suggests that the more restaurants we see in a given area, the more prosperous it is.[4] That seems to have held in the distant past as well.

Poehler, who has worked with Ellis at Pompeii for almost two decades, said that there's been a shift in how archaeologists understand the way middlers transformed the urban design of Pompeii. A century ago, he said, scholars fretted about how places like Stephanus' fullery were signs that Roman culture was in decline. They assumed that Pompeii's noble, cultured aristocrats were being pushed out of towns like Pompeii by dirty, low-class traders, and this shift in turn led to the demise of civility. This theory was inspired partly by Victorian prejudices against working-class people, especially at a time when most archaeologists would have been upper class. But it also came from reading what the Romans themselves had to say. Petronius' *Satyrica*, a novel-like description of Rome's underbelly written during Nero's reign, features a long description of the tasteless parties thrown by the *libertus* Trimalchio, who indulges in *Great Gatsby* levels of vulgar conspicuous consumption. Nearly all the descriptions we have of middler life were written by elites like Petronius, and a lot of it is disparaging.

Poehler took a swig of beer and laughed with Ellis. Today's archaeologists are a lot more skeptical of fictionalized tales like *Satyrica*, which probably reflected prejudices more accurately than realities. Instead, he and Ellis view this period as one of renewal, when opportunities for middlers shifted the balance of power.

But how do you prove that *liberti* and other middlers were not monsters who destroyed the empire when so few of them left any records of what they thought? There are no eloquent refutations of Petronius' snide remarks about Trimalchio. Even monuments to middler power like Eumachia's building are ambiguous because we know so little about how it was used. To re-create middlers' lives, Ellis and Poehler turned to a new method of historical investigation called data archaeology. Through careful observation, they aggregate information about many structures and objects—like, say, hundreds of bars—to figure out the typical habits of individuals. It's the perfect method for exploring a lost way of public life.

Gutter data

"At Pompeii, archaeology tends to look for the big, the mighty, and the unusual," Ellis said. He was referring to the villas and monumental buildings that have been the subject of so many excavations. "But what we do is look for the usual. I'm looking for the most common events that happened on the streets. Eric [Poehler] looks for what's happening in the street."

He doesn't mean that metaphorically. Poehler is the author of a book called *The Traffic Systems of Pompeii*,[5] and his research involved spending a lot of time literally squatting in the street, analyzing stones that were once covered in manure, sewer runoff, and carts. Lots and lots of carts. So many, in fact, that most streets in Pompeii are scored by two deep groove marks where carriage wheels wore down the stones.

This immediately tells us something important: cart sizes, or at least the space between wheels in the chassis, were relatively standardized. And that in turn suggests a widely accepted set of social norms for driving in cities.

Crawling around in the gutters, Poehler also noticed distinctive, wedge-shaped bites taken out of curbs at intersections. After counting them, noting their positions, and then consulting with an engineer, he figured out what had caused them: thousands of poorly executed right turns from the right lane, as carriage wheels banged into the curbs or rode up onto them. There were no signs of similar wear on the left-hand sides of the intersections, suggesting left turns went wide, the way they do on US streets today. Even a terrible driver making a left-hand turn wouldn't smack into the left curb. These clues strongly suggested that people in Pompeii drove on the right side of the street.

I stood at an intersection on Via Nocera, a block from Via dell'Abbondanza, imagining carts crowding past me, while people mobbed the tabernas. The streets in Pompeii are deep, with high curbs; to cross them, I jumped across three large, flattened boulders that served as crosswalks. Poehler said they were built like stepping stones in part because the streets often flowed with dirty water. I hopped from stone to stone, trying to imagine the scene with a sewage-laced river gushing below. A cart rode up on the curb, splashing us with filth as it slid back down, and the air filled with curses in Latin, Punic, and Oscan. This is the kind of moment that Poehler and Ellis are conjuring with data archaeology, and in that instant it made Pompeii's past feel more tangible to me than knowing where emperors walked and consuls lived.

The crosswalks themselves are another hint that carts were standardized, since the stones are perfectly spaced for those two wheel ruts to pass between them. There are also hints from literature at the time that carts may have been permitted in the city only at night, when pedestrian traffic was thinner. Some Roman municipal regulations also stipulated that carts weren't allowed downtown on feast days, so that those revelers from the Temple of Isis wouldn't be run over by traffic roaring up via Stabiana.

In a sense, data archaeology represents the democratization of history. It's about looking at what the masses did, and trying to reconstruct their social and even psychological lives. Ellis has used data to peel back our preconceptions about Roman life and reveal a thriving group of middlers who loved to shop and eat out. When Ellis looks at a city, he said, he sees it as a "volumetric matrix" of building materials and human labor. "I always wonder: How does that volume come to be there?" he mused. He's asking, quite literally, what it took to move giant heaps of stuff around in the ancient world. The answer leads back to the absence I mentioned earlier, the blank space that remains where *liberti* and slaves' perspectives would have been, if we were standing on Via dell'Abbondanza two millennia ago.

Poehler argues that we can learn a lot from absences, which makes sense for a person who reconstructed traffic in Pompeii by studying what had been worn away from rocks. "I'm interested in the part of the rock that is now gone," he told me. "The shape that's worn away—that's what people did." This is especially true when it comes to public spaces where many people were doing roughly the same kinds of things. "If you take the hundred thousand interactions with the stone in aggregate, all over the city, the absence is thousands of people making the same decision. Now, suddenly, you have a picture of a system of traffic at a place like Pompeii where we had zero evidence ever before." Poehler paused, and I thought about all the absences in my home city that mark the places where crowds gather: bald trails worn through park grass; the dings in subway paint where commuters have whacked their bags into the walls repeatedly; and yes, the scars in streets where cars took turns too quickly, or bounced as they hit the bottom of one of San Francisco's many steep hills. In these nicks and cracks, Poehler believes we catch a glimpse of the anonymous masses whose lives have been lost to history.

We can even see how the social stratification that shaped people's lives is written into the cityscape. For Romans, paved streets were a key technology, making it easier for carriages to deliver goods and people, as well as making it more pleasant to walk around. But the city's rich didn't pay to bring this fancy technology to everyone. Large, posh streets like Via dell'Abbondanza were paved of course, as were most main thoroughfares. But in the poorer, eastern parts of the city, many streets were made from dirt.

After the earthquake, roads in the western side of the city near the temple district were quickly put to rights, while side roads leading to humble residences were not. Poehler described a "swanky residential area" in the northwest where every street is paved with stone, except for two that were paved cheaply, with beaten earth and ash. The unpaved streets serviced the back doors of villas, and the front doors of inexpensive homes. "This tells us that some people in the city control the application of this [paving] technology, and they're not going to share it with everybody," Poehler said. The message to the lower-class people on that street was clear: they were only good enough to share a street with the back sides of wealthier people's homes. Pompeii's paving system may seem like a wonky detail of urban infrastructure, but it tells us a lot about how neighbors treated each other in Roman cities.

Rise of the *liberti*

It's only recently that researchers have figured out what percentage of the Roman population were *liberti*. Henrik Mouritsen estimated their numbers after conducting a painstaking overview of sources that compiled every name mentioned on a gravestone in Rome and a few other key locations in the empire. Buried in these lists was a predictable data pattern. Roman patrons preferred to give their slaves foreign names, particularly Greek ones, as a way of emphasizing slaves' otherness and inferiority. Even when slaves were freed, their slave names followed them for life. Roman government officials used a specific nomenclature for designating a *libertus,* incorporating his former master's last name into his new freedman name. Sometimes public records even include an "L" for *libertus* next to the freedman's name to make it abundantly clear that this person was once property. To find the graves of *liberti,* Mouritsen's sources counted these telltale Ls, as well as Greek and foreign first names that had Roman last names attached.

Scholars also sussed out the *liberti* population by examining government documents. Emperor Augustus passed a law that kicked *liberti* off the grain dole,[6] and that allows scholars to count how much the dole roster shrank when *liberti* were excised. Based on those numbers, and grave markers, researchers extrapolate that up to three-quarters of free people in cities were either ex-slaves or their descendants. Sandra Joshel, author of *Slaves in Rome,* used similar methods to measure slave populations, which she pegs at roughly 30 percent of all people living in cities. Obviously we can't arrive at any exact numbers, especially because we have so few records of the slave and *liberti* groups. But there is no denying that Roman slavery and manumission were common, and *liberti* represented the new face of Roman urbanism in the first century.

We can even see distinctly *liberti* anxieties reshaping architecture at Pompeii. Painfully aware of negative stereotypes about them, these new middlers often tried to build houses that could blend into rich neighborhoods by appearing to be more sumptuous than they actually were. One strategy, which I saw at the House of the Tragic Poet, was to create an illusion of more space inside. The owners of this famously well-decorated house wanted neighbors to think they had a large peristyle garden surrounded by marble columns, the way Julia Felix did. So they placed a few columns at a cunning angle in their modest garden, making it appear to passersby on the street that they were seeing a little slice of a much bigger peristyle. These faux villa decorations seem like a direct forebearer of the way people today make small rooms look bigger by using tall mirrors and brightly painted accent walls.

Some middlers tried to assimilate into the upper classes by adopting their elevated tastes. One of the most evocative and haunting examples of this is a fresco from a stately villa called the House of Terentius Neo on Via Stabiana. (Its name comes not from the owners, but from an election poster outside that urges passersby to vote for Terentius Neo.) In the painting, a married couple stand together, posed exactly like two aristocrats would be in family portraits seen throughout the city's elite villas. And yet the pair are defiantly middlers. The man, a baker, is dressed in a citizen's toga, which means he's neither aristocrat nor slave. The woman holds a pen and wax writing tablet, which are the tools of a bookkeeper. This likely identified her as a *liberta*, since bookkeeping was a common job for female slaves. There's something beautiful and defiant about this couple who chose to be painted in an elite style, but not to hide the signs of their former servitude. It was a subtle but powerful way of asserting that *liberti* were as good as their freeborn counterparts.

Other *liberti* dealt with their class anxiety by reveling in ostentatious displays outside their homes, as if to thumb their noses at the haters. Nowhere is this more obvious than at the House of the Vettii. A pair of *liberti* wine merchants, likely brothers, the Vettii had a palatial villa in the fashionable northwestern region of the city. Next to their front door, they painted a picture of Priapus weighing his giant erection on a comically tilted scale. Keep in mind that Priapus' generously sized member was a symbol of wealth in ancient Rome, and male nudity was not considered obscene. Despite all that, this painting would have read as low-class satire, a dirty joke for gutters rather than peristyles. It was as if the brothers wanted to remind their aristocratic neighbors of their low-class origins—and their incredible financial success in spite of it.[7]

On the prowl for more signs of *liberti* life, I met archaeologist Sophie Hay on Via Stabiana, several blocks south of the House of the Vettii. Hay spent years excavating a nearby taberna and adjoining villa that belonged to a *libertus* named Amarantus. The day was hot, and the noontime sunlight was brutal. When Hay arrived, her shoulder-length blond hair in slight disarray, she was parched. We hunkered down together at the edge of the road like two ancient Romans, sharing the bottle of cold water that I'd filled from the mouth of a cupid on one of Pompeii's many restored fountains. As we talked, Hay's story brought to life the working-class neighborhood that had been here two millennia ago.

If you turned right onto Via dell'Abbondanza from Via Stabia, then turned right again onto a narrow side street called Citarista, you'd find Amarantus' bar a little ways down. When the *liberti* set up shop there, the centuries-old villa attached to his bar was badly in need of repair, thanks to years of neglect and the recent earthquake. Amarantus wanted to give it a posh do-over, but apparently he preferred to do it in the cheapest way possible. Artisans restored a beautiful dining room in the back of the house, complete with frescoes. But when it came to what Amarantus did with his atrium, Hay didn't mince words. "The roof is crappy," she said. "It was just reeds held together with lime." The impluvium was little more than moldings on the floor in the shape of a pool that couldn't actually hold water. In fact, most of the atrium served as a dusty storage area for dozens of amphorae filled with wine for the taberna. Amarantus kept the previously high-class bedrooms off the atrium intact, but he housed his mules and dogs there; Hay herself excavated the animals' remains where they lay when the ash fell. Perhaps Amarantus was imitating the style of the fullery and bakery up the street, where workshops filled the atria once used by local luminaries to make business deals or political maneuvers. But the bar owner apparently also wanted his house to retain some of its former aura as a villa. Otherwise, why bother building a fake impluvium into the middle of his warehouse space?

Whatever his pretensions to upward mobility, Amarantus' customers were likely other *liberti* or middlers like himself. "It was probably not a posh bar," Hay said with a grin. "It was for artisans and people laboring in workshops. The neighborhood has at least two or three paint shops and a garum manufacturer. There's a lot of mercantile activity going on, and there's another bar opposite to him. They are probably feeding and watering the locals." Hay's colleagues have figured out some of the menu items from the bar, too. By analyzing food remains from the taberna cooking pots, and in a giant pile of what Amarantus' customers left behind in his toilet cesspit, they've determined that Amarantus' taberna offered staple items like fish, nuts, and figs. The food, Hay says, was plentiful and high quality. Pompeii may have been shaken by class divisions, but the farms surrounding it were bountiful enough to feed rich and poor. Amarantus was also experimenting with imported wine; among his 60 amphorae of Cretan wine, Hay found one amphora of wine from Gaza. "It's the only wine from Gaza ever found at Pompeii," she marveled. "I like to think he was trying to offer his customers something a little different."

Amarantus was also, like many of his *liberti* neighbors, participating in local politics. Archaeologists first discovered Amarantus' name—a Latinized version of a Greek name, appropriate for a former slave—in a hand-painted sign outside his shop that urged patrons to vote for his preferred candidate. Unfortunately, the painter misspelled both Amarantus' name and the name of his candidate. "Maybe the guy who painted the sign was a bit drunk," Hay suggested. There was a painter's shop next to Amarantus' place, and she speculated that maybe Amarantus had traded some wine for his neighbor's skill with the brush. The results may not have been an artistic masterpiece, but they offer strong evidence that Amarantus was knitted into the political fabric of his city—just like the baker and his wife who urged people to vote for Terentius Neo. Sure, Amarantus spent most of his time working at the bar. But he also had passionate opinions about how his middler customers should cast their votes.

Queen of the Cocksuckers

Seven blocks north of Amarantus' bar, on a shady side street near the city walls, astute observers out for a stroll would find a very different electoral suggestion. There, someone scrawled misspelled graffiti that reads: "Isadorum aed / optimus cun lincet." A rough translation would be: "I implore you to vote Isadorus for Aedile / he licks cunt the best."[8] It was surely a backhanded compliment. Though perhaps Isadorus felt a flush of pride at being named for his sexual prowess, Romans generally considered performing oral sex a lowly task for slaves and women. But this satirical election notice is hardly unusual; Pompeii is saturated with sexual graffiti and imagery. Archaeologists excavating the city in the 18th and 19th centuries were shocked to discover loads of erotic paintings displayed on the walls of fine houses, and pictures of disembodied penises decorating public squares, shopfronts, and even sidewalks. The fertility god Priapus and his astonishingly large penis weren't found exclusively at the House of the Vettii, though that's a particularly memorable rendition. Priapus was a popular icon all over Pompeii. The city is perhaps as well known for its dirty pictures as for its importance as an archaeological treasure.

But all these dick pics are in fact part of what makes Pompeii an archaeological treasure in the first place. They are perhaps the most jarring example, for modern Westerners, of the radical cultural disjunction between pre-Christian Roman culture and what came after. For the people of Pompeii, the Vettii brothers' Priapus painting would have been immediately legible as a rambunctious way of signaling their financial success. Penis-shaped wind chimes and carvings were considered lucky, and many shops had them on display for the same reason that shopkeepers today put those cute waving cats (*maneki-neko*) in their windows. There was not much taboo on sexual imagery in ancient Rome, reflecting a culture that didn't treat sex and sexual organs as the forbidden subjects they would become in the Christian world.

Despite changing attitudes toward sex in the late 20th and early 21st centuries, the sexual objects found at Pompeii and neighboring Herculaneum are still kept in a special "Secret Cabinet" area of the Naples Museum. There, curious students of history can gaze in awe at baskets full of clay dicks, or admire charming penis figurines with feet, wings, and their own little penises (yes, they are penises with penises, because you can never have enough luck). Plus, there are elegant statues of gods humping various animals and people.

The lure of this forbidden history is what draws so many visitors to Pompeii's brothel, called a lupanar (she-wolves' den). It's a plain, triangular two-story building located at an intersection in Amarantus' neighborhood off Via dell'Abbondanza. The lupanar was likely as sensational two millennia ago as it is now, but for very different reasons. Today, of course, tourists who were forced to learn Latin in school are titillated by the idea that the supposedly great men who shaped our culture were also boinking in a place with smutty pictures on the walls and built-in plaster beds. Back in Amarantus' day, it would have been a special treat to buy sex at what University of Washington archaeologist Sarah Levin-Richardson calls a "purpose-built brothel."[9] The reason she uses the phrase "purpose-built" is to underscore that it was a specialty retail outlet. Horny Romans could buy sex pretty much anywhere entertainment was to be found, and sex workers typically worked out of rooms in tabernas or shop fronts built into their masters' villas. Others worked the streets around busy areas like the Forum. Devoting an entire establishment to sex work would have been unusual, sort of like a restaurant serving only food made with chocolate. It was quite simply an unusual establishment. That's probably why the lupanar at Pompeii is the only purpose-built brothel that archaeologists have yet uncovered in the Roman world.

The day I visited the lupanar, it was the most crowded attraction in the park. A steady stream of tourists entered the front door on one street, passed quickly through a hallway punctuated by entrances to rooms with built-in beds, and exited just as quickly through another door onto the next street. Directed by tour guides speaking Italian, Japanese, and English, they glanced up at the erotic frescoes painted in a series of panels above the doorways, where men and women cavorted in various groups and positions. It looked a bit like the brick-and-mortar version of a gateway to an adult website: Click here for threesomes; Click here for gay male; Click here for doggy style. As someone who grew up with internet porn, the faded images of half-dressed figures on pillow-strewn beds struck me as relatively tame, like a sex comedy set in a college dorm. Though today the place feels airy and open, during its heyday many of the rooms were cramped and dark.

It was common for *liberti* to become sex workers, but some of the workers here were slaves with no choice about what they did. Still, Levin-Richardson has uncovered evidence that the women of the lupanar were not in some horrific *Handmaid's Tale* dystopia. Many took rebellious pride in their work. Levin-Richardson spent years studying the city's purpose-built brothel for clues about what its employees were like. She found some of her answers in dirty graffiti, much like the fake election poster about "cunt licking." Though it was long believed that the plentiful graffiti in the lupanar was written by men, Levin-Richardson points out that a great deal of it is clearly authored by women. Female literacy was common in Pompeii, and literate slaves helped their masters keep household accounts like the *liberta* in the portrait at the House of Terentius Neo. At least a few of the sex workers here had to be literate because she found graffiti written by a person who identifies herself as female. One simple sentence, written on the lupanar's wall, reads: "fututa sum hic." It means "I (a woman) was fucked here."[10]

Other graffiti are boastful claims from women about their sexual prowess. Several women identify themselves as "fellatrix" or "fellatris," which is a female noun version of the verb "to suck." One possible translation might be "queen of the cocksuckers." Particularly fascinating is a phrase written in the brothel's hallway: "Murtis · Felatris." The stylized letters, complete with middle dot, imitate the way names and titles of prominent men were written on the Forum walls. Murtis, Queen of the Cocksuckers, wrote her name in the same fashion a *Rector provinciae* would, turning her marginal role as a prostitute into something as exalted as a governor. Other women took on the title "fututrix," converting the verb *futuere* (to fuck in the active role, to penetrate) into a noun that could be translated as "fucktress" or "fuckmistress." Women who called themselves fututrix were not just playing with the idea of a political title like Murtis did. They were also claiming a dominant social role. In Roman culture, men made a strong distinction between the penetrating and penetrated person during sex; the penetrated person was viewed as low status, like women or slaves. As a fututrix, the woman was the penetrator, and thus her client was subordinate.

I stepped out of the flow of people streaming through the lupanar and into one of the rooms with its now-bare plaster bed. In the 70s CE, this place was piled with blankets and pillows, lit by lamps, and full of freshly painted graffiti proclaiming its occupants to be as elite as the men who owned them. By looking beyond the writings of rich men, into the back streets and slave quarters, we find evidence for a society where rigid Roman social roles were being literally rewritten from the bottom up. Former slaves like Amarantus and the Vettii brothers could achieve wealth and influence. Women like Julia Felix could be property owners. And the names of sex workers like Murtis would be remembered for thousands of years, while the names of her clients burned to dust.

And yet, despite over two centuries of researchers excavating Pompeii, very few people understood the world inhabited by Murtis and Amarantus until recently. Partly that's because data archaeology has given us new tools to explore the lives of nonelites. But it's also due to a more fundamental problem with the way we study history. Though people of the 19th and 20th centuries treasured Pompeii, returning to it repeatedly for further excavation, there were parts of its culture they wanted to forget. When they came upon sculptures of genitalia or dirty graffiti, they locked these things away in "secret cabinets" because it was too hard to step outside their Christian values and look at those artifacts with Roman eyes. Only in 2000 was the "secret cabinet" in the Naples Museum opened to the general public. Roman sexuality is

so alien to modern people's sensibilities in the West that it was practically illegible. Museum curators in previous centuries treated lucky
penis charms like pornography, and historians didn't consider prosti-
tutes worthy of study.

But turning away from understanding this part of Roman culture
prevents us from fully appreciating the social fabric of a place like
Pompeii, where privates were quite literally public.

Roman toilet etiquette

I barely glanced at the arches and pedestals in the Forum. I was hunting for an unmarked room to the northeast of this hallowed hall for
elite politicians. I finally found it, identifiable only by a single window placed far above eye level in its high wall. Inside, a trough along
the wall was choked with dirt and weeds. It was one of the city's few
public bathrooms, and its design was as jarring as seeing a disembodied penis painted next to a shopkeeper's door. Today it's difficult
to make out the shape of the toilets in what was once a rather dark,
enclosed space with that one high window to let out the stench. But
with the help of Olga Koloski-Ostrow, a Brandeis University classicist who has written an in-depth study of Pompeii's sewers,[11] I was
able to piece it together. Along one wall ran a deep trench that was
once full of gushing water headed for the sewer. A few stone blocks
jutting from the wall marked where there would have been a bench
with several evenly spaced U-shaped portals for the great men of the
Forum to lift their togas and relieve themselves. "There's about 30
centimeters between seats," Koloski-Ostrow told me. "It was pretty
standardized. Unless you're very fat, you're not rubbing thighs with
the guy next door."

Still, there were none of the privacy barriers between toilets that we expect in bathrooms today. Everybody was sitting nearly cheek-to-cheek on a bench. And there was even less personal space when it came to toilet paper. When a Forum visitor was finished doing his business—these public toilets were largely reserved for men—he would grab a stick with a sponge on the end called a xylospongium, dip it in a shallow gutter of running water at his feet, thread it through a hole in the bench below his bum, and wipe. In public and private toilets, xylospongia were shared.

Often it is in the most squalid and filthiest of places that we can uncover profound truths about a society that considers itself civilized. In the toilets of the Forum it's obvious that Roman moral authorities weren't obsessed with covering up body parts or bodily functions, the way Christians were. Instead, they focused on controlling how people moved through urban spaces. As Koloski-Ostrow put it to me, the Forum toilets weren't really about modesty. "I'm sure many Romans defecated on the streets, in alleys, and outside the city walls," she said. "We have graffiti on the edges of the city saying 'Don't defecate here,' and you wouldn't put that there if people weren't doing it." Public toilets, she said, were about controlling behavior. "The elite Romans build [toilets] where they do because they don't want human excrement on the Forum floor. They don't care about the streets, but they want a pristine look to the polished imperial Fora. It's a way of regulating space, of saying, 'This is where you're going to do your business.'"

The more I spoke with Pompeii experts, the more I heard comments about how Romans wanted to "regulate" space. From the streets to the tabernas, every public area was caught in a web of formal and informal rules. Even in the lupanar, graffiti reflects a society deeply concerned with the social meaning of sexual positions.

There's a symbolic link between Roman selfhood and the physical organization of people within cities. Unlike the residents of Çatalhöyük, who were in the early stages of developing emotional and political entanglement with the land, Roman urbanites were born into a world where sedentary life had eclipsed nomadism thousands of years before. Over time, most of the crafts and activities done in the home at Çatalhöyük had exploded outward, becoming public places throughout the city: bakeries, fulleries, cemeteries, temples, jewelers, sculptors, painters, tabernas, and yes, toilets. The city was less an agglomeration of homes than a resplendent, complex public space. People's homes were largely public, with atriums open to the street and serving as receiving areas for business associates and guests. This trend only intensified as middlers converted their homes into live-work spaces where the line between business and private life was thin at best. You might say that Romans expressed their entanglement with the land by dividing up their cities into specialized public zones devoted to everything from sex and defecation, to amusement, political activity, and bathing. Moving into and between these spaces was a way of being a Pompeiian.

If we pull out for a wide-angle view, we might apply this same notion to the entire Roman Empire. Each city had its own specialized function, or its own role to play in the greater glory of this sprawling civilization that had wrapped its arms around the Mediterranean. Pompeii was a town for revelry, renowned for its beauty and tasty food. It was the naughty but beloved stepdaughter to the stately, powerful city of Rome. When it was lost in a moment of uncontrollable, terrifying violence, it caused an historical trauma that went beyond the horror of losing thousands of lives. Public spaces had been destroyed, and with them a part of Roman identity. Rome's reaction to the eruption of Vesuvius was therefore not the long, slow detachment we saw at Çatalhöyük. Nobody decided to abandon Pompeii. Its fiery burial was felt as an almost unbearable loss—and the many survivors hastened to rebuild their lives in other cities, devoting themselves to constructing new versions of the public spaces they had lost.

CHAPTER 6

After the Mountain Burned

I t started with an earthquake. People living in cities around the Bay of Naples were used to quakes, however, and the shockwaves they felt that day in the autumn of 79[1] probably didn't alarm anyone very much. They continued to run their businesses, deal with the harvest, and spout off in the Forum. But then Vesuvius started to smoke. Nobody in the Roman world had recorded stories about volcanic eruptions before, and people writing about Vesuvius in Latin later described the mountain covered in "a black and dreadful cloud, broken with rapid, zigzag flashes, [and] behind it variously shaped masses of flame: these last were like sheet-lightning, but much larger."[2] There were no easy words to describe what must have seemed like an unimaginable disaster. Smoke filled the sky for at least a day, and maybe two, before the mountain began to launch rocks—some as big as the stones that paved Pompeii's more upscale streets.

The earthquakes continued. At that point, people started to panic and leave the city. In wagons and on foot, people gathered up their valuables and fled north or inland as rocks rained down on rooftops, smashing walls and cracking ceramic shingles. We have only one eyewitness account of escaping the eruption, from Pliny the Younger, who recorded his experiences decades after the events that killed his uncle and thousands of other people in Herculaneum and Pompeii. He describes smoke filling the air as he and his aunt evacuated along with crowds of people, the darkness so profound that they stumbled frequently.

Despite the obvious danger, we know that thousands of people stayed behind. Free people remained by choice, while slaves remained under orders from their masters. As a meter of ash and rocks accumulated in the streets, even the holdouts must have realized it was time to go. Sophie Hay, who told the story of Amarantus, said that the Pompeii we find beneath the ash is a city in disarray. People had packed up their valuables, and moved their possessions into more protected places. "Nothing is where it's supposed to be," Hay said. And everyone was moving. More than half of the city's population died in the streets[3] fleeing through the city's southern neighborhoods as ash and mud flowed in from the north.

Perhaps the most poignant record of people's last minutes can be found outside Pompeii, on the docks at the wealthy enclave of Herculaneum. There, in the warehouse rooms typically used for loading and unloading cargo, archaeologists uncovered dozens of bodies. Herculaneum was located much closer to Vesuvius, to the north of the volcano, and death came more swiftly there. Skeletons are pressed together in tangled heaps at the back of the warehouses, many clutching sacks of valuables. These are the charred remains of people who waited for rescue boats that never came. It's easy to imagine the horror they endured, cowering from liquid flames erupting from the beautiful, green mountain that had formed the backdrop for so many of their garden parties and festivals. These people may have lived in linens, and ate reclining while servants brought them wine, but they died in lowly storehouses like slaves. Their would-be rescuers died too; Pliny the Younger says his uncle perished after piloting his ship to help with rescue efforts.

In all, 1,150 bodies have been found in Pompeii. Given the likelihood that more bodies will be found in non-excavated portions of the city, archaeologists generally estimate that a tenth of the city's population of 12,000 perished.

The final blow at Pompeii came not from the rain of rocks and ash, but from what geologists call a pyroclastic flow, or several blasts of superheated gas that instantly cooked every living thing in their paths for up to ten kilometers around Vesuvius. After this scouring surge, ash continued to pour from the sky, burying Pompeii beneath six meters of hot, toxic material. As the bodies of humans, horses, dogs, and other animals decayed beneath the ash, they left hollows behind. In the 1860s, archaeologist Giuseppe Fiorelli was able to pump plaster inside those hollows to re-create the positions and even facial expressions of the volcano's victims. Visitors to Pompeii today pass through two massive display cases of these plaster bodies as they enter the park next to the amphitheater. They're grisly and affecting—it's hard not to think of them as corpses, rather than the casts of bodies reduced to dust long ago. Some people anticipated death, with arms raised defensively over their heads, while others slept peacefully. It made me think about the afterlife of Çatalhöyük. Centuries after Çatalhöyük's abandonment, people living on the Konya Plain used the place as a burial ground for their dead, and considered the land holy.

Pompeii, too, has become a monument to the dead. Though we find signs of life everywhere in its streets and shops, there is no way to visit this city without confronting how horrifically it was extinguished. For the Romans who lived through 79, this feeling was far more intense. The disaster shook the entire empire, and its refugees poured into nearby cities, forever haunted by the violent loss of their home. Perhaps because it was so terrifying, the eruption became an event that people seemed to want to erase from history. When I asked street expert Eric Poehler about it, he marveled at how almost nothing is said about such a major event in the Roman world. But he said it became less mysterious to him after he learned about the idea, taken from 20th-century history, of a "generation of silence" that comes in the wake of disaster. A similar kind of cultural silence followed the Spanish Flu pandemic of 1919, which slaughtered over 675,000 Americans in a matter of months—more than had died during all of World War I. Despite the widespread devastation caused by the disease, governments and media downplayed its severity. And after the pandemic was over, almost nobody wrote about it.[4]

The Romans' silence about the destruction of Pompeii can be read as a measure of how traumatizing the eruption was. Unlike the many fires that devastated Rome, and the wars that had pummeled the Republic, this was a disaster that could not be solved with money or manpower.

"An absolute nightmare"

When I first began researching the abandonment of Pompeii, I was perplexed by how suddenly people gave up on the city. In 79, the Roman Empire was at the height of its wealth and influence. Why didn't Emperor Titus send a bunch of slaves down to dig Pompeii and Herculaneum out from under the ash? Certainly it would be a huge job, I thought, but the city of Rome was famous for rebuilding itself after multiple devastating fires. That was an enormous undertaking, and so was building the aqueducts. It wasn't like Titus was afraid to spend money. During his father Vespasian's reign, he poured an enormous amount of resources into the sacking of Judea. Then he spent his first year as emperor finishing construction of the insanely expensive Colosseum project his father had started. Engineers constructed the Colosseum to hold water so that Romans could witness mock naval battles. Given how complicated it was to build something like that, why wouldn't Titus want to show his power by rebuilding Pompeii, too?

One standard answer to this question is that people were afraid to go back to Pompeii, terrified of supernatural forces that caused the earth to spit fire. But Romans were a lot more pragmatic than that. People who had survived the devastating earthquake in 62 returned to their homes to rebuild, and middlers took the opportunity to transform abandoned villas into shops. So it wasn't as if disasters emanating from the earth had stopped Pompeiians before. This only deepened the mystery for me. I started to wonder whether this wasn't simply a case of Pompeii being a low priority for the Roman elites. Though it was a treasured resort town a century before, the empire had expanded so much that pleasure-seekers could plan beach vacations along the coasts of Spain and Portugal. Cities in North Africa were gentrifying, with Roman-style urban grids in Carthage and neighboring Utica blotting out traditional Punic layouts.[5] This also meant better sources for garum, the local fish sauce delicacy that was one of Pompeii's major exports. Maybe Pompeii had simply fallen out of fashion, or had become a political annoyance? It seemed to me that Titus and the Roman elites had made a calculated decision that Pompeii simply wasn't important enough to merit a concerted recovery effort.

And then I talked to Janine Krippner, an expert in pyroclastic flows, who teaches geology at Concordia University in New Zealand. She's studied the eruption of Mount St. Helens in Washington firsthand,

and has also visited other regions after disastrous volcanic eruptions that are similar to the one that wrecked Pompeii. When I asked her by phone what would have happened after the Vesuvius eruption, she was emphatic. "It would have been a living hell, and it would have gone on for years," she said. "Rehabilitation would take generations. It would have been an absolute nightmare." She quickly answered my main question, which was why people couldn't have dug Pompeii out. "New snow has the density of 50–70 kilograms per meter cubed. Ash has the density of 700–3,200 kilograms per meter cubed. The sheer work to dig that city out without bulldozers would have been enormous." She paused, thinking. "On top of that, the flows would have been hot for a long time." The temperature of the mud and ash flows started out at 340°C, and would have retained that heat thanks to the insulating layers of rock and ash above them. Plus, the ash itself would have emitted toxic fumes and particulates. Anyone working in those conditions would suffer from the extreme heat while inhaling volcanic ash that rapidly made them sick.

The disaster went beyond the city's walls, however. This was an environmental catastrophe that affected the entire Naples Bay region. Krippner pointed out that the waterways feeding Pompeii would have been clogged with mounds of toxic ash, cutting off freshwater supplies and the transit network that linked the coastal city with its neighbors. And then there were the long-term effects to the land. Krippner compared the Vesuvius aftermath to the Mount St. Helens eruption, where there is almost nothing growing nearby even after 40 years. When the wind blows, ash still swirls into the air, creating noxious gusts. Pompeii was known for its fertile farmland and delicious food, and Vesuvius would have snuffed that out in an instant—even if people were able to remove the ash after things cooled down. "That amount of volcanic ash can prevent soil from getting oxygen, and it can cause acidity in the soil, too," Krippner explained. "That reduces the avail-ability of nutrients for crops, so now you're struggling to get anything to grow afterwards." The eruption that killed thousands at Pompeii and Herculaneum had also sterilized the soil for kilometers around. It had literally poisoned the land.

Natural disaster tore the city out of its residents' hands without warning. They desperately wanted to go back, but couldn't. Emperor Titus himself toured the city's smoking ruins, looking for ways to mitigate the damage.[6] There was no way to do it. Even with modern technology, the task would have been insurmountable. But they managed to survive, and bring Pompeii's memory with them into their new lives. The fate of Pompeii gives us a chance to see what happens when people are forced to abandon a city against their wishes. In the last few years, scholars have found evidence for massive resettlement of refugees throughout the region, as well as new building projects in nearby cities like Naples and Cumae, where the streets filled with former Pompeiians trying to start again.

The luck of Gaius Sulpicius Faustus

Naples is a noisy city, full of narrow cobblestone streets that roar with cars and motorcycles careening uphill from the Bay of Naples at terrifying speeds. These downtown roads were built for the kinds of mule-drawn carts that dominated the ancient and medieval Roman worlds, but now pedestrians fight for space alongside metal machines that Murtis and her friends at the lupanar could only dream of. Still, a lot hasn't changed. The walls are covered in epic amounts of graffiti, and the bars are off the hook.

Back in 79 CE, when this city was called Neapolis and its sidewalks swirled with ash from the volcano, refugees from Pompeii began to trickle in. Some brought carts and sacks full of valuables; others arrived with nothing but soot in the folds of their robes. Many would have been sick from inhaling the volcanic particles that Krippner described, coughing and vomiting and weak from walking the long road from Pompeii for a couple of days. Some fled here because they had family who could take them in, and others because it was the only nearby town they knew. We can't be certain what happened in the immediate days after the disaster, but it's likely that refugees would have overwhelmed the city's inns. New arrivals might have slept outside. Temples and amphitheaters would have opened their doors to shelter terrified people who had lost everything. It would have been a scene familiar to anyone who has seen the aftermath of hurricanes and wildfires today.

What might surprise us is how similar the Roman government's response was to what we hope for in Western democracies of the early 21st century. Emperor Titus toured the disaster sites, and subsequently offered survivors financial support to rebuild their lives. Suetonius, who published a biography of Titus in the early 120s, explains: "Immediately [Titus] chose commissioners by lot from among the ex-consuls for the restoration of Campania; the property of those extinguished by Vesuvius, and who had no surviving heirs, he donated to the restoration of the affected cities." Miami University classics scholar Steven Tuck, who has conducted groundbreaking research on the survivors of Pompeii, said that the "restoration of Campania" referred to building what appear to be entirely new neighborhoods for refugees in several coastal cities, including new temples dedicated to popular Pompeii gods like Venus, Isis, and Vulcan, as well as baths and amphitheaters. Partly this money would have come out of Rome's coffers, but Suetonius also says that it came from the "property of those extinguished by

Vesuvius." Given how many exceptionally wealthy people had vacation homes at Herculaneum and Pompeii, we have to assume that this was quite a windfall.

Tuck tracked the pathways of survivors to Neapolis, Cumae, Puteoli (now Pozzuoli), and Ostia, using some of the same techniques that scholars have used to identify *liberti*: he examined grave markers. When last names and tribal names exclusive to Pompeii start appearing on grave markers in other cities, that indicates a refugee population. Thanks to Tuck's sleuthing, we know that survivors in Neapolis included the Vettii, whose Pompeii shop was decorated with that memorable painting of Priapus weighing a phallus as big as his entire body. We can't be sure the brothers who owned that shop were the survivors, as they were likely just two of the many *liberti* connected with the family. But at least some of the extended Vettii clan made it to Neapolis, and tried to stick by other disaster survivors. Intermarriage between refugee families was quite common, suggesting that survivors probably lived alongside each other and continued to share many things in common. L. Vettius Sabinus, a Vettii, commemorated his wife, Calidia Nominanta, on a tomb inscription—her name, too, is one that was found exclusively at Pompeii before the eruption. Another tomb at Neapolis commemorates Vettia Sabina, whose husband left an inscription that contains an Oscan word from the original language of Pompeii.

Most of the survivors we know about, Tuck said, are *liberti*. He thinks that's partly because families fled Pompeii together, including their *liberti* and slaves. But he also suspects *liberti* survived because many of them were out of town on business when Vesuvius erupted. It was common for *liberti* to continue working for their former owners after being set free, and typically they took jobs managing their patrons' financial interests and agricultural lands outside Pompeii.

Tuck said this kind of work arrangement would also explain why so many Pompeii refugees preferred to resettle in cities along the northern coast of the Bay of Naples. This wasn't just out of convenience. Rich Pompeiian patrons had business assets there, and *liberti* could continue managing them after their masters had perished.

Tuck has a favorite Pompeii survivor: a man named Gaius Sulpicius Faustus, a freed slave from a family of bankers who lived at Pompeii. Gaius and the Sulpicii left behind the kind of paper trail that historians dream of discovering. We know they made it out of the city because they dumped a strongbox as they fled, full of records that tied the Sulpicii to a small financial empire that included several warehouses at Puteoli. These were the exact kinds of holdings that a *liberti* like Gaius might have managed for his patron. In 79, Puteoli was the main harbor in ancient Italy, where large cargo ships unloaded their bulk commodities like marble, timber, grain, and wine. The Sulpicii would have warehoused these items before sending them up to Rome on smaller boats. Gaius' trail picks up again in the pretty seaside town of Cumae, where Tuck discovered graves bearing the names of several Sulpicii *liberti*. Tuck figures that Gaius and his family must have continued managing his master's holdings after the disaster, settling in Cumae because it reminded them of Pompeii. Like Pompeii, Cumae was a commuter town for people who had business in Puteoli. Tuck noted that this pattern was so common because Puteoli was basically a warehouse town, and not a particularly nice place to live.

The Sulpicii weren't the only family with this idea. Tuck has found evidence that Titus, and later his brother Emperor Domitian, funded the construction of entire new neighborhoods at Cumae to house refugees, complete with baths, an amphitheater, and temples dedicated to Pompeii's patron gods Venus and Vulcan. Plus, he commissioned a brand-new road linking the city to the Roman road network. Perhaps not surprisingly, the new neighborhood had a meeting house for Augustales, the *liberti*'s association. "This wasn't slave labor that got shipped in, either," Tuck said. "It meant jobs for local people." The road was an especially luxurious addition to the city, making it more accessible for trade and tourism from Rome. At Puteoli, the emperor commissioned an amphitheater that was a perfect copy of the Colosseum in Rome. "[The survivors] are getting state-of-the-art facilities," Tuck marveled. "It's extraordinary and unprecedented. I imagine people looked at [that amphitheater] and said, 'We've got it as good as Rome.'"

Though thousands died at Pompeii, it appeared that the Roman government smoothed the way for life to go on for the tens of thousands of refugees in Campania. We can't assume that everyone got an equal piece of the pie, as most of our records come from the lives of wealthy *liberti*. But refugees remained together in their new homes, intermarrying and often carrying on the same businesses they had in Pompeii. Few people wrote publicly about the trauma, but they held on to their identities as Pompeiians.

There was one thing they did discard within one generation, though: their *liberti* status. All the children of *liberti* from Pompeii stopped using their parents' slave names, so nobody in Cumae or Neapolis or Puteoli would know that they were descended from slaves. Instead, the public would know them only as Vettii or Sulpicii, venerable Pompeii families whose fortunes continued to grow with the empire.

The movement of people from Pompeii to Cumae and Neapolis was very different from the journey back to village life from Çatalhöyük. Though some of Çatalhöyük's residents probably migrated to other mega-sites like Domuztepe with its Death Pit, most of them rejected high-density urban life in favor of smaller communities. Pompeiians sought out cities very similar to the one they had lost, and Vesuvius' refugees were able to maintain continuity in their lives despite losing almost everything. This was largely because Rome had colonized the entire area, creating public spaces that were in some ways interchangeable.

What Pompeiians lost was their city's hybrid culture, where traditions kept by the native Oscan-speakers mixed with new ideas from Egypt, Carthage, Rome, and dozens of others. Still, as the story of the Sulpicii family makes clear, international trade on the Mediterranean continued. And the city's middlers gained more social standing than they might have in Pompeii. They shed the memories of slavery, and what their parents endured so their children could be freeborn. It was a determined forgetting that paralleled the way all of Rome tried to forget what happened when the burning ash fell.

On my last evening in Pompeii, I took a stroll through the city for a few hours before sunset. It's astonishing how much walking through Pompeii today replicates the experience of living there 2,000 years ago. The streets are crowded with families speaking many languages; children yelp and leap across the fat blocks of the crosswalks; and hot, tired people dunk their heads under the street fountains for relief. It's easy to imagine the bustling place it once was, full of the smell of sizzling meat, the tang of spilled wine, and the reek of fermented fish sauce—mingled with the stench of the streets, which must have been an unpleasant soup of garbage, wastewater, and poop from every animal in the city (including humans). I made my way from the villas ringing the Forum down Via dell'Abbondanza, the crowds thinning as the sun sank. At last I stood almost alone on a corner near Julia Felix's insula.

I took pictures of the chipped marble on a taberna counter as a tourist filled her water bottle from one of the restored public fountains across the street. Built from rough blocks worn smooth by hands, the waist-high square tub caught a steady stream of clean, cold water pouring from the mouth of a pipe. This ancient piece of urban infrastructure went back thousands of years, but was deceptively simple. Its existence depended on a sophisticated idea of public space, and an economic system that provided the rocks, pipes, and city planning. And all that was supported by a political hierarchy that assigned people to different roles according to written and unwritten rules: merchant, slave, aristocrat, wife, patron, prostitute. On the streets of Pompeii we can find a record of those roles changing, but also remaining intact at some fundamental level, like a Roman road that has lasted for millennia beneath six meters of volcanic stone.

PART THREE

Angkor

THE RESERVOIR

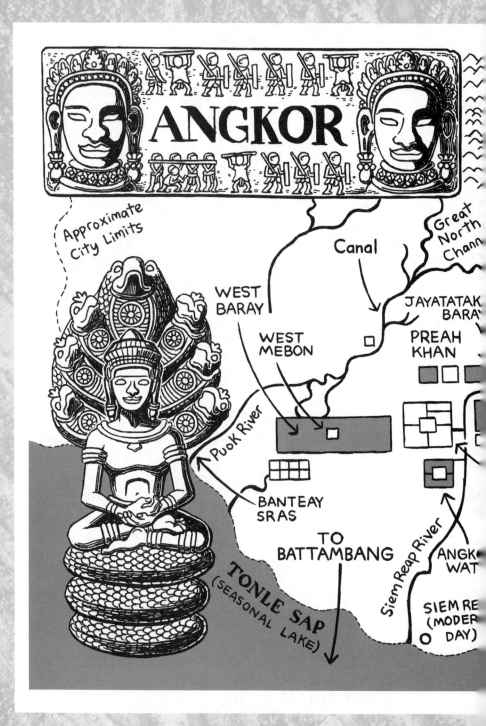

ANGKOR

Approximate City Limits

Great North Chann

Canal

WEST BARAY

WEST MEBON

JAYATATAK BARA

PREAH KHAN

Puok River

BANTEAY SRAS

TO BATTAMBANG

Siem Reap River

ANGK WAT

TONLE SAP (SEASONAL LAKE)

SIEM RE (MODER DAY)

ANGKOR THOM

To Preah Khan →

ROYAL PALACE

WEST GATE

VICTORY GATE

EAST GATE

BAYON

CANAL

SIEM REAP RIVER

SOUTH GATE

1 Km

TO BENG MEALEA AND KOH KER

EAST MEBON

EAST BARAY

Kulen Mountains

SRAS SRANG

CHAU SREI VIBOL

NGKOR HOM

Reservoir

Roluos River

Approximate City Limits

N

IHARALAYA (RE-ANGKOR CAPITAL)

ROLUOS (MODERN DAY)

TO SAMBOR PREI KUK

10 Km

CHAPTER 7

An Alternate History of Agriculture

When I arrived in Phnom Penh during Cambodia's dry season in January, I stumbled through the streets in a jet-lagged daze, barely seeing the dense city around me. My mind was on thousand-year-old Khmer temples, their golden facades crumbling into worn stone blocks and imprisoned by thickly braided tree roots. These structures, from the Khmer Empire's capital at Angkor, have been synonymous with the myth of lost cities for at least two centuries. You can even find Lara Croft exploring the legendary ruins of the Angkorian temple Ta Prohm in the first *Tomb Raider* movie. But unlike Roman civilization, Khmer traditions are not lost or dead. The culture that blossomed at Angkor—a form of Theravada Buddhism combined with centralized state power—continues to shape many aspects of Cambodian life today. Once I'd gotten some sleep, I could see it on the streets of Phnom Penh, the city where Khmer royals fled in the 15th century as Angkor fell apart. Today, the nearly 600-year-old capital's buildings are obscured by tangles of electrical cables instead of tree roots, and fences around modern-day palaces are topped with coiled razor wire so fine it shimmers in the sun like jewels.

Phnom Penh is joined to Angkor by the Tonle Sap River, which winds north from the modern city before widening into the Tonle Sap Lake that provided the ancient capital's farms with nourishing floodwaters every year. Eleven hundred years ago, Angkor was one of the biggest metropolises in the world, thronging with nearly a million residents, tourists, and pilgrims. When the 13th-century Chinese diplomat Zhou Daguan visited, he described elaborate city walls, breathtaking statues, golden palaces, and vast reservoirs with artificial islands. And yet even as Zhou fought his way through crowded streets to witness the king's sumptuous processions, the city was pregnant with its own demise. The Khmer kings were losing their hold over the empire's provincial capitals abroad, and neglecting the city's crucial water infrastructure at home. Some years, Angkor's dams burst during rainy season; other years, silt choked the canals and slowed the flow of mountain water to a trickle. And each time this happened, repairs got harder. Farming got harder. Trade slowed down, and political tensions heated up. By the mid-15th century, the city's population had fallen from hundreds of thousands to mere hundreds.

Though obvious in retrospect, it was the kind of incremental catastrophe that nobody could recognize until it was too late. That's what makes Angkor's abandonment so haunting. On a day-to-day timescale, people living there wouldn't necessarily have noticed the city's dramatic transformation. There was no giant sign proclaiming the end of life as they'd known it; instead, there was a mounting pile of annoyances and disappointments. Nobody was fixing the canals, and the reservoirs were flooding. Some of the once-thriving neighborhoods had fallen empty and silent. There were no more fun parades on festival days. Younger generations would realize they had fewer economic opportunities than their elders had. In the 14th century, an Angkorian kid with talent might grow up to become a full-time musician or scholar at court. Or she might have a thriving business selling spices on the heavily trafficked roads to temples at Angkor Wat and Angkor Thom. But by the late 15th century, young Angkorians had few choices. Mostly they grew up to be farmers. Some became priests or monks, tending what remained of the faded temples.

In the soft apocalypse at Angkor, we can see directly what happens when political instability meets climate catastrophe. It looks chillingly similar to what cities are enduring in the contemporary world. But in the dramatic history of the Khmer culture's coalescence and survival, we can see something equally powerful: human resilience in the face of profound hardship.

Jungle farming

Somehow, Angkor managed to exist at a size bigger than many modern cities for hundreds of years, despite the fact that this region of Cambodia is known for its climate extremes, with rainy season floods and dry season droughts. While their kings waged wars abroad and fought internecine battles in court at home, the Khmer people razed the tropical jungle and replaced it with an orderly city grid, complete with elevated, flood-proof houses and a canal network for drinking water and irrigation. The Khmer built towns, hospitals, and bureaucracies at a rate that would have made Rome's emperors jealous. How did this medieval civilization thrive in an environment that would be challenging for us even today?

The answer, say archaeologists, is not that the Khmer were somehow ahead of their time, nor that they were in league with ancient aliens. (*Of course* there are people who claim that Angkor was built by aliens.[1]) Instead, it was because Khmer urbanites came from a tradition of tropical city-building that looks very different from what we see in the more northerly regions of the Levant and Europe. For nearly 45,000 years, the Khmer's ancestors were perfecting the techniques required to build and farm in the jungle, manipulating earth and water to make empires whose remains often melted back into nature, leaving very little trace.

It probably began with a forest fire. Fifty thousand years ago, humans in Southeast Asia were fanning out across the South Pacific in reed boats, eventually island-hopping all the way down to Australia. During that time, they settled in the lands that would one day belong to the Khmer Empire, as well as on islands that we know today as Indonesia, Singapore, the Philippines, and New Guinea. In all these places, bands of roving humans foraged at the edge of dense tropical jungles, living on plants and small animals. At some point, they would have noticed that forest fires had a paradoxical effect. Though initially deadly, the flames also cleared away underbrush and left a layer of charcoal behind. Some of humans' most beloved foods, like yams and taro, flourished after the jungle had been torched—partly because they had more room to grow, but partly because those charred bits created a more nutrient-rich soil. After observing the benefits of wildfire, says Max Planck Institute archaeologist Patrick Roberts, humans figured out that they could start these fires themselves and reap the benefits.

Roberts is the author of *Tropical Forests in Prehistory, History, and Modernity*,[2] a fascinating survey of how the equatorial jungles incubated civilizations that looked very different from the ones like Çatalhöyük in the Levant. In areas as far-flung as Southeast Asia and the Amazon, Roberts and his colleagues have found clear evidence that humans set off controlled burns. Sometimes they would work the soil with their hands afterward, mixing in the charcoal along with animal bones and feces to create more fertile ground. Over thousands of years, they learned how to encourage certain trees and plants to grow, scattering seeds from banana trees, sago palms, taro, and other starchy staples, and eventually changing the tree populations of the forests where they foraged. When they paddled between islands, they brought their seeds and burning techniques with them, carrying favored plants and small mammals back to Southeast Asia. From South Asia, they brought chickens down to the South Pacific Islands in boats as well. It wasn't agriculture exactly—it was more like proto-farming. The groups doing this were probably still nomadic. But even millennia later, scientists can use stratigraphic techniques to see the ways these ancient people altered the jungle. Lower (older) layers are packed with a mix of fossilized pollen and seeds from a naturally occurring mix of plants, but upper layers are full of remains from plants that are noticeably skewed toward crops favored by humans.

While people were molding bricks to make the first houses at Çatalhöyük, people across the world in the highlands of New Guinea were digging deep trenches to drain a swamp known today as Kuk. The people of Kuk Swamp built elaborate structures to live in, and planted bananas, sugar cane, and taro in the drained farmland they'd created. Their settlement was the culmination of generations of humans working the earth. A landmark paper published by Roberts and his colleagues in 2017 in the scholarly journal *Nature Plants* sums up: "There is now clear evidence for the use of tropical forests by [humans] in Borneo and Melanesia by c. 45,000 years ago; in South Asia by c. 36,000 years ago; and in South America by c. 13,000 years ago."[3] By the time we reach the Angkorian period, people in Southeast Asia would have had plenty of experience building settlements in an extreme environment.

Roberts says this doesn't mean tropical urbanites somehow "beat" more northerly communities in the race to build cities. "Clearly, urbanism is different in different parts of the world," he told me. "We need to be more flexible in how we define this." Cities aren't made of the same materials throughout the world, nor do they have the same design. Roberts continued, "The tropics demonstrate that where we draw the lines of agriculture and urbanism can be very difficult to determine." This has sometimes made it hard for archaeologists to identify urban remains that aren't as recognizable as stone walls and figurines. To find early cities in Southeast Asia, scientists look for what they call "anthropogenic geomorphology." (To break down all those Greek roots, *anthropogenic* means human-caused and *geomorphology* means the study of earth-shaping.) The term encompasses all the ways that humans have sculpted the land for their own uses, from planting trees and mixing fertilizer into the soil to draining swamps and building artificial hills as foundations for wooden huts.

Understanding the ancient origins of anthropogenic geomorphology is key for recognizing the remains of cities like Angkor, where only a tiny fraction of the urban grid was made from stone. The cities that arose out of the long history of tropical agriculture weren't high-density clusters of stone buildings ringed by farms like Çatalhöyük or Pompeii. Instead, they were low-density sprawl that incorporated large swatches of farmland into the urban fabric. Homes and public dwellings were made from earth and perishable plant materials. Though impressive, these urban features quickly returned to wilderness after people had abandoned them. When European archaeologists first visited Angkor, they were conditioned to look for Western modes of urban development, and thus the vast majority of homes in the city remained invisible to them. They made a beeline for the stone towers of Angkor Wat and Angkor Thom, mistaking these temple complexes for small walled cities, instead of walled compounds within a massive urban sprawl. They completely missed the once-packed neighborhoods, reservoirs, and farms that had left their marks on the land for acres around.

Everything is better with lasers

It became very obvious to me how archaeologists could make that mistake when I visited Sambor Prei Kuk, once the crowded capital city of the seventh-century Chenla Empire in Cambodia. Now all I could see were a few scattered temple towers and a 1,300-year-old wall that looked like a hill covered with underbrush. Sitting on a broad rock and looking around, I couldn't imagine these crumbling structures as part of a metropolis. And yet Sambor Prei Kuk, with its Hindu temples and large reservoir, was in many ways the prototype for Angkor. Shaded by the trees that give this place its name—Sambor Prei Kuk is Khmer for "the temple in the richness of the forest"—I pored over maps of the area with archaeologist Damian Evans. "There was a huge wooden city here once," Evans said, waving his arm in the direction of a small dirt road paved in fallen leaves. "Once it rotted away, what remained were moats and ramparts and mounds." That's what he's got on his map, which shows the ground elevations around us in granular detail.

Evans and his colleagues created this map and many others of the Angkor region using an imaging technology called lidar, short for "light detection and ranging." Lidar instruments scatter laser light off the planet's surface, capturing the photons as they bounce back up. By analyzing the light pattern with specialized software, mapmakers can re-create ground elevations down to the centimeter. Lidar is ideal for studying anthropogenic geomorphology because the rain of light slips between leaves, peeling away the forest cover to reveal the city grid that once was. With funding from the National Geographic Society and European Research Council, Evans coordinated a team that conducted broad lidar surveys of Angkor in 2012 and 2015. The system may have been high tech, but it was also DIY. Their mapping rig started with a Leica ALS70 HP lidar instrument, roughly the size and heft of two portable generators. Operators mounted the lidar inside a protective plastic pod and then attached the whole rig to the right skid of a helicopter. Secured next to it was an off-the-shelf digital camera taking pictures of everything, so they could match up the lidar data with regular old photos. The system was effective, but a little awkward for passengers. "We had to rip out most of the seats in the helicopter to put in a power supply and hard drives," Evans recalled. But the discomfort was worth it. What they found has helped rewrite the global history of cities.

Evans and his colleagues' lidar maps resolved a longstanding mystery about Angkor and its environs. For centuries, archaeologists and historians had been perplexed by inscriptions on Angkorian temples that suggested the city's population was close to a million people. That would make its size equal to the largest cities in the world at the time, competing with ancient Rome at its height. It seemed impossible, based on the remains they could see of Angkor Wat and Angkor Thom. How could so many people have stuffed themselves inside those walled enclosures? Nineteenth-century Western scholars were loath to believe an Asian city could achieve such stature, and later researchers were skeptical about the accuracy of inscriptions ordered by the king. It wasn't until Evans and his team revealed the landscape in and around Angkor with lidar that it became obvious that those inscriptions were no exaggeration. Today, Evans argues that the population was likely 800,000 or 900,000, making Angkor one of the world's biggest cities in its heyday. After demonstrating how much lidar could reveal, researchers used the technique to look at other parts of the Khmer Empire, too.

One of those places was Sambor Prei Kuk, the city that predated Angkor's rise, where I was poring over a lidar map with Evans. I quickly discovered how disorienting it was to compare what a machine could see with lasers, and what I saw with my eyes. Around me were leafy trees and rolling hills. But on the map, I could see a late-700s urban plan: elevation measurements revealed thousands of square and rectangular mounds that once served as the foundations for temples and houses. The rocks where we'd stopped for a lunch break were in the city center, surrounded by a near-perfect square where a now-eroded wall once stood tall, possibly edged by a moat. Depressions in the ground that I had taken for natural swales were actually the remains of deep reservoirs and canals. Peering more closely at the map, I noticed hundreds of tiny mounds like goosebumps around the temples.

"What are those?" I asked Evans, imagining some kind of specialized agricultural feature.

"Termite mounds," he replied, pointing to a lump of earth nearby. "They love it at this elevation."

Not everything the lidar sees is from a vanished civilization. But those termite mounds were a reminder of how powerful the technology is—it can pick out extremely small features in the landscape—and how deft the researchers are at identifying the difference between ancient structures and natural features of the modern forest. Trying to ignore the insect cities overrunning the land around us, I returned to contemplating the works of humanity. Elevated causeways led away from the temple entrances and stretched out into the Tonle Sap, forming long fingers of earth that are still visible in the shimmering water. At Sambor Prei Kuk, kings of the Chenla Empire worshiped the Hindu god Shiva, unlike the Angkorian kings who preferred Vishnu.

One of the most striking temple towers here is an octagon of deep orange sandstone. A flying palace is carved into one wall, its soaring towers and balconies borne on the backs of birds. Inscriptions here and in the other temple remains testify to the glory of these Hindu kings, but very little is written after an inscription about the first Angkorian king, Jayavarman II, declaring himself a divine leader in 802. At that point, Angkor began to rise and Sambor Prei Kuk slowly emptied out.

Still, Sambor Prei Kuk remains an important place for the Khmer to this day. In one temple, we found fresh baskets of incense, paper flowers, and a golden parasol sheltering a statue of the Buddha. But the centuries-old Buddha was also a modern touch. It had been built on top of an ancient lingam shrine that symbolizes the power of the Hindu god Shiva. Linga, which are found in temples throughout the Khmer Empire, can take many forms, but most often they are square pedestals with a smooth, abstract phallus shape—the lingam—mounted straight up in the middle. A stylized moat surrounds the lingam, connected to a narrow spout that juts out from the lip of the dais. This is sometimes called the yoni. Priests would pour liquid offerings over the lingam, allowing it to run into the moat before spilling out of the spout. It was an evocation of fertility, a reenactment of the way life-giving water flows down from mountain stone. Especially for people living in the river valley, where runoff from the Kulen Mountains fed the land, this would have been a powerful image.

I considered Evans' lidar map showing the square walls around Sambor Prei Kuk's downtown, with the temple's earthen features spilling out into the Tonle Sap. It looked like an enormous version of the lingam shrine. As I made my way through temples and city centers across the Angkorian Empire, I saw this pattern of squares and waterways repeated on various scales, from diminutive linga to the enormous square moats nested around Angkor Thom.

But Evans was less interested in the perfection of the city's cosmological design than he was in the commoners' neighborhoods that lie beyond the temple enclosure's walls. Outside, he noted, "there is no rigid urban grid," though the lidar map offers plenty of evidence that thousands of people lived and farmed there. Architecture historian Spiro Kostof argues that all city layouts can be grouped into two basic types: organic and grid.[4] Organic city plans are ad hoc, with winding roads and ever-changing improvised structures like the ones at Çatalhöyük or in many medieval European cities. Then there are cities built on grids, like most Roman ones, whose growth is often regulated by a centralized government. Cities in the Angkorian tradition exhibit both patterns, often with a strict grid surrounded by organic forms. These organic Angkorian neighborhoods often belonged to people who built the city and provided food for its inhabitants. Their histories did not register on Western archaeology's radar until Evans and his colleagues used a literal radar device to call attention to them.

The city before the city

Angkor's remains are located today next to Siem Reap, a thriving cosmopolitan city that hugs Tonle Sap Lake. Like the modern-day town of Pompei, Siem Reap conjures crowds similar to those that would have come to see the sights centuries ago. A festive atmosphere pervades the tourists' quarters in town. Shops offer tourists "happy pizza" spiked with dried cannabis, and tuk tuk drivers pull up along the sidewalks, offering rides to temples and nightclubs. Road trippers buy petrol for their scooters from vendors selling it by the liter in recycled alcohol bottles. Local hipsters and students hang out at Brown's Coffee, a chain that's like an upscale Starbucks with much tastier drinks and snacks. The items for sale may have changed since Angkor's heyday a millennium ago, but the energy hasn't. In the city and the temples beyond, you'll hear a cacophony of languages from across the Eurasian continent. It's easy to believe that for more than a thousand years, people have come here to witness the glory of Angkorian civilization.

It wasn't always like that.

In the early days of Angkor's construction, the city's eventual ascendancy was by no means assured. University of Hawaii archaeologist Miriam Stark has been excavating around Angkor for most of her career, and she's interested in the city's humble, nascent stages. She and I spoke by video shortly before she left for the summer excavation season in 2019. Lounging at her kitchen table in Honolulu, she talked casually about how she'd avoided the Khmer Rouge while excavating in Cambodia during the mid-1990s. Sharp and funny, Stark projects a restless energy when she explains Angkor's history.

Like many pre-Angkorian villages scattered across the region, the early communities along the northern Tonle Sap Lake were centered around earthen mounds with wooden shrines on top. "Angkor is like one of those shrines on steroids," she laughed. She's right. If you can swallow your awe at the city's breathtaking temples, you'll notice that they are essentially outsized, ornate shrines atop earthen platforms. The city that sprawled around these temples was also built by people who remolded the earth to create foundations, roads, and pools for their homes. This is a tradition that goes back to Sambor Prei Kuk, but also much further, to distant Pleistocene ancestors who burned and churned the soil.

Stark sees the rise of Angkor as a spiritual process more than a feat of urban planning. "People were attracted to religion," she mused. "And to spectacle. There's a way in which you get intoxicated by ritual and practice." She believes that people were initially attracted to the region because they were visiting the local temples and shamans. When Jayavarman II declared himself the first Khmer king, it was in a religious ceremony in the Kulen Mountains. He continued his state-building very near the place where Angkor would rise, founding a city called Hariharalaya. (Today, archaeologists call it Roluos.) There, Jayavarman II built temples and reservoirs, and held massive festivals and rituals. Though cities grow by promising people wealth and security, Stark says we can't dismiss the lure of entertainment that would have gone along with Jayavarman II's religious displays. Angkor began as a metropolis founded on pageantry and political spectacle.

Recently, Stark and her colleague, University of Oregon anthropologist Alison Carter, have been excavating domestic houses in Battambang, a province south of the Tonle Sap. Essentially it was a suburb of Angkor. There, they have found settlements dating back thousands of years, which means its residents witnessed the birth of Angkor from across the seasonally swollen lake. "We say Angkor began in 802," Carter told me, referring to the date when King Jayavarman II claimed the land that later became Angkor. "But when did people in Battambang think Angkor began? I wonder what they were thinking about what was happening across the lake." The question is a good one because we know Angkor was occupied long before King Jayavarman II came to town. The villages in Battambang had their own leaders, complete with inscriptions chronicling their deeds, so it wasn't as if they were simple farmers waiting for a god-king to tell them what to do. They must have greeted the swelling metropolis with a mixture of curiosity and dread.

Some historians trace Angkor's early culture to influences from India, where both Hinduism and Buddhism originated before flowing into Southeast Asia. Jayavarman II explicitly wanted to build a Hindu

empire. Inscriptions carved after his death recount a coronation ceremony where he declared himself the Khmer's godlike ruler in a ritual
that borrowed concepts of divine kingship from Hindu traditions. But
Stark and Carter think the picture is a lot more complicated than
a sudden infusion of Indian Hinduism. "It's not Indianization—it's
globalization," Carter said, noting that influences came from many
parts of Asia. "Plus," she added, "by the time Angkor arises, there's
a thousand years of indigenous cultural development in Cambodia."
The local people in places like Battambang were just as important to
Angkor's development as ideas from abroad.

There is one transitional moment in Khmer history that stands
out to archaeologists. In the centuries before Jayavarman II consoli-
dated the region under his banner, people stopped burying their dead.
Southeast Asian settlements from roughly 500 BCE to 500 CE are
full of burials, with all the attendant artifacts that archaeologists rely
on to take the measure of the culture they're studying. But after the
late first millennium, there are virtually none. Bodies may have been
cremated, or taken outside the city to be picked clean in the jungle.
Scholars attribute this change in burial practices to the rise of Hinduism and Buddhism, but it could have come from other traditions, too.
Like Angkor's demise, the city's origin was so complex and gradual
that there's no easy way to demarcate its beginning.

In terms of population, however, in the ninth century the land
across the lake from Battambang starts to get crowded. University of
Hawaii anthropology researcher Piphal Heng told me that there are
two basic theories about what drew people to the land that Jayavarman
II claimed. The first takes us back to the way people had been molding the earth for millennia. Heng emphasized that the longstanding
communities in the area all had a similar layout, with homes clustered together and rice fields stretching out around them. "What this

meant was that the entire city, other than its core, could host both settlements and rice fields," Heng said. Having a city full of rice fields offered a twofold advantage. Obviously, it meant more food for the non-farming elites and their households. It also meant that the city would be more like Los Angeles–style sprawl than Manhattan-style density. And that gave Angkor's leaders a strategic advantage over the enemies lurking at every border. "They could control the land farther away, down to the lake or to the north and northwest," Heng pointed out. A city whose architecture includes farms is quite simply bigger and more imposing, uniting people over much bigger distances than a densely packed city like Çatalhöyük could.

There's a second theory about why people came to Angkor, however, and it's a lot harder to measure than hectares of farmland. As more people arrived, there was an opportunity for the elites to mobilize enough workers to create and maintain the city's water infrastructure. Pre-Angkorian cities generally featured large reservoirs called barays to store water during the dry season, so this was a continuation of a long tradition—only on a gargantuan scale. To keep its rice fields soaked year-round, Angkor would need the biggest barays and canal networks the world had ever seen. At that moment in the ninth century, we see the start of a self-reinforcing cycle: Angkor's swelling population required water storage, but the water storage system couldn't be maintained without enormous labor forces. The city had to keep growing to slake its thirst.

Throughout the city's lifetime, its water systems were more than a pragmatic way to keep the rice farms going. They were also monuments where the city's rituals took place. Pilgrims visiting the city's temples would travel there by boat, across human-made reservoirs and moats. Angkor Wat was dedicated to the Hindu god Vishnu, who is pictured in one of the temple's most famous reliefs in the middle of a dramatic battle between gods and demons. This battle, a tug-of-war where the rope is a huge snake, churns up a sea of milk. Vishnu intervenes and liberates the universe from demonic control. This became an origin story for the Khmer people, which is why many of Angkor's greatest artworks depict Vishnu floating on a sea of milk, orchestrating the birth of the world. One of Angkor Wat's most famous monuments was a six-meter-long bronze statue of Vishnu reclining on one of his four arms. He's resting in the middle of a square pool, surrounded by a square artificial island, in the middle of the rectangular West Baray reservoir. This is the same island I described in the introduction, where Evans griped about how cosmological designs don't always translate well into good water engineering.

Like all massive urban infrastructure projects, the Angkorian canal and reservoir system failed repeatedly and spectacularly. It's a cautionary tale about how cites can create and destroy ecosystems. At the same time, labor politics at Angkor were also an ecosystem, and as we'll see, it was a pretty delicate one.

CHAPTER 8

Empire of Water

When the Chinese diplomat Zhou Daguan visited Angkor in the late 13th century, he marveled at the city's weather. "For six months, the land has rain, for six months no rain at all," he wrote. "From the fourth to the ninth month it rains every day." Today, meteorologists would say that Cambodia is buffeted by two different monsoon systems. From May through October, the southwest monsoon brings heavy rains from the Gulf of Thailand and the Indian Ocean. The Tonle Sap River would overflow its banks into Angkor's rice fields, leaving only the tips of trees visible over its roaring waters. Then, from November through March, the northeast monsoon howls down out of the Himalayas, soaking parts of India but trapping Southeast Asia under a unique low-pressure zone known as a monsoon trough. Though the edges of this trough often stir up violent tropical storms, at its center the weather becomes intensely hot and dry. Caught between these two powerful monsoon forces, Cambodia seesaws between climate extremes. For Angkor to sustain its population of nearly a million, the Khmer had to build a social system based on the regulation of water.

Debt slaves and their patrons

In the early 900s, roughly a century after Jayavarman II declared himself the divine king, King Yasovarman moved the capital slightly northeast and directed Angkorians to dig an enormous reservoir known as the East Baray. Emperors generally built a baray to celebrate their ascension to the throne,[1] but this one was different. For one thing, it was enormous. Measuring 7.5 kilometers by 1.8 kilometers, the East Baray was a long rectangle that held roughly 50 million cubic meters of water—the equivalent of 20,000 Olympic swimming pools. To fill it, workers built a canal that redirected the Siem Reap River into the center of Angkor.

The East Baray would have dominated the city center's low-density mix of temple mounds, wooden houses on stilts, and rice fields spread along the western banks of the Tonle Sap. These were early days for the metropolis; it would be two centuries before workers cut the sandstone to build the dramatic towers of Angkor Wat. Yasovarman had to order entire neighborhoods of people to abandon their homes to make way for his monumental project. He also would have had to muster the equivalent of an army to build the thing. And that was another way the East Baray was different from previous reservoirs. It was one of the first Angkorian infrastructure projects to require vast amounts of human labor drawn from all over the growing empire.

When I visited, the East Baray had melted back into the jungle, and time had smoothed its earthen retaining walls into a gently sloping landscape thick with trees and scattered farms. It's hard to believe this place was once a luxurious ceremonial center where Yasovarman led his retinue through lush pageants. Maybe that's the point. Only a huge labor force could have turned this wild land into a symmetrical pool, and now that they're gone, so is the baray. The true marvel of Angkor was its workers. And yet we rarely hear about them in historical accounts or inscriptions from the walls of Angkorian temples. They are the anonymous masses executing the will of Yasovarman.

Every time I interviewed an archaeologist about Angkor's city plan, I always asked who built the barays. Thinking of Rome, I imagined that it had to be slaves. But their answers were complicated because there isn't an easy equivalence between the way labor was organized in the Roman and Khmer Empires. Angkorian inscriptions suggest that kings and other elites kept workers, but a common Old Khmer word used for this group, *khñum*, can refer to a wide range of roles.[2] *Khñum* could mean temple workers who were lifetime slaves, often taken from ethnic minorities (Zhou calls them "savages") or imprisoned during war.[3] They could also be indentured workers, sometimes referred to as debt slaves,[4] who endured temporary bondage as a form of taxation. Sometimes these workers are listed in temple inscriptions as property, alongside other valuables like fabrics, precious metals, and animals. *Khñum*, like Roman slaves, ran the gamut from manual laborers to learned scholars. These workers are also identified by many titles, including *gho, gval, tai, lap,* and *si,* which are used to mean everything from "worker" and "servant," to "slave" and "commoner." For simplicity's sake, I'll refer to them as *khñum*.

The *khñum* debt slavery scenario sounds brutal until you consider that most capitalist cultures in the West use a similar system. In the United States, it's not unusual for people to graduate from college with so much debt that they have to work their whole lives to pay it off. Others take on debt to pay for a house or buy a car. Though technically all of us can choose what kind of work we do to pay off these debts, it's rare to find anyone who is doing the exact kind of work they'd like to do. Many of us feel like we're being told to dig ditches by some distant corporate authority, or risk losing everything. Still, we keep working instead of rising up against the banks, for complicated reasons. Maybe we don't want to rock the boat because our lives are relatively comfortable, or maybe we need health insurance to pay for a child's hospitalization, or maybe the corporations seem too powerful to defeat. Those feelings might have kept *khñum* in line, too.

Angkorian society was built on debt slavery, but the idea of indebtedness permeated every layer of public life. Inscriptions on temple and palace walls reveal that everyone in Khmer society owed something to someone.[5] Even Khmer kings owed their subjects clean water, roads, and other amenities. Debt also cemented political connections between the rulers of outlying kingdoms through the Khmer patronage system. Distant elites paid Yasovarman in precious metals, fine fabrics, and more ineffable tributes like favorable trade relationships and supplies of human labor. In return, Yasovarman gave them huge amounts of land to farm. If farming didn't strike a royal's fancy, Yasovarman could bring him to court to enjoy the pleasures of the city. There are records of some kings bestowing weird ceremonial positions at the Angkorian court, including fan-bearer, barber, and wardrobe keeper.[6] Presumably these were well-compensated sinecures that gave his allies an excuse to hang out in the Angkorian court and goof off.

Yasovarman did more than share his spoils with his fellow aristocrats, though. He and other kings frequently left Angkor, making perilous journeys to visit the courts of their subjects. Ostensibly they did this to be worshipped, but it was also a way of acknowledging the importance of their people. The king may have been powerful, but he was nothing without all the labor power that turned Angkor into a shining cosmopolis.

Labor was Angkor's most valuable asset. That's not because the Khmer had an unsophisticated economy; indeed, slave-owning societies all the way up into the 19th century often depended largely on their workers to generate wealth. Sociologist Matthew Desmond, writing about slave labor in the US South, has noted that by the time the Civil War started, "the combined value of enslaved people exceeded that of all the railroads and factories in the nation."[7] The Khmer Empire was held together by a system that normalized servitude by casting it as something people owed to their leaders, and by incorporating it into public rituals. As Miriam Stark puts it, "Leaders cajoled more than they coerced, and used display as much as military might to legitimize their rule."[8] Still, Angkor's many attractions only existed because its people felt obliged to build them. A king who gave them nothing would eventually get nothing in return.

Stark emphasized how unstable this whole arrangement was, partly because it depended on loyalty at every level of the social ladder. At the top was the king with his family. Directly beneath them were other noble families who lived in Angkor alongside ministers, officials, and a hereditary priest class who served as advisers to the king. And then, out in the provinces and countryside, there were semi-autonomous local systems of government. Ruling alongside the king's inspectors and local officials were usually village chiefs and a council of village elders. And beneath them all was the largest group: the khñum, made up of slaves, commoners, and servants. To expand the empire, the king needed all these groups. And because there were no rules about succession, even people at the top could fall—and those beneath them could rise. This led to wars over succession, local uprisings, and recurring cycles of chaos. Pondering the city's eventual demise, Stark mused, "Suppose what collapsed was as much social as it was environmental or physical?"[9]

The urban population explosion

Over a hundred years after the construction of the East Baray, a new Angkorian king won a long battle over succession. In the early 11th century, King Suryavarman became Angkor's first expansionist king, growing the Khmer Empire's borders north into Laos and Thailand, and south to the Mekong River Delta in Vietnam. Partly he accomplished this thanks to his strong relationship with the Chola Kingdom in what is now southern India. The Chola were a source of war allies and trade throughout his reign. But Suryavarman's success as a king was also due to his relentless focus on city-building. During Suryavarman's reign, new cities rose along the Tonle Sap River, as well as the Mekong, Sen, and Mun, all of which were natural rivers that radiated outward from the Angkor area. The king was growing an urban network connected by water, which could be used for the transit of dignitaries, as well as trade. According to Ball State historian Kenneth Hall, the number of places in the Khmer Empire using the Sanskrit word "pura" as a suffix, meaning "city," jumped to 47 during Suryavarman's reign. Fifty years earlier, only 12 cities were recorded.[10] Suryavarman's *khñum* constructed roads and temples in far-flung regions, sometimes leaving a linga shrine behind as a sign of his sovereignty.

At Angkor, Suryavarman's passion for urbanism expressed itself in perhaps his best-known monument: the West Baray, still recognized today as one of the largest reservoirs ever built without the aid of industrial equipment. Located a few kilometers from the East Baray, it would turn the king's palace into a jewel set between two great, long artificial seas. Looking at Evans' lidar maps of the city, it's easy to see that the long rectangles of West and East Baray line up neatly along an east-west axis, though the West Baray is taller. An even greater showcase of labor power than the East Baray, the West Baray measures approximately 8 kilometers by 2.1 kilometers. A person ambling along its banks at a leisurely pace could spend an hour walking from one end to the other, and circumnavigating both reservoirs might take a whole afternoon. To keep the West Baray full, workers shoveled out another canal to divert water from the Siem Reap River, which was already feeding the East Baray; they also built canals connecting it to the Tonle Sap. River water was supplemented by rain during the monsoon seasons. At its most flooded, the West Baray probably held about 57 million cubic meters of water,[11] putting its size at roughly 23,000 Olympic pools. Construction took so long that it wasn't finished until after Suryavarman's death in the mid-tenth century. The reservoir is partly full again today, thanks to 20th-century reconstructions.

Even when the West Baray was completed, the work of maintaining it and the rest of the city's water infrastructure would have been ongoing. We have to imagine that Suryavarman used his patronage of many distant kingdoms to relocate thousands of people from the country to do this work. Some would have been sent by their local rulers as tribute to the king, while others did it to pay their taxes. Building the West Baray was one way to make sure that the precarious political hierarchy was stable.

It was also Suryavarman's way of using urban design for revisionist history. The *khñum* digging the reservoir had to raze all the neighborhoods, roads, and farms built around Yasovarman's old palace enclosure, probably evicting residents in the process. And, as they reached the deepest levels of the reservoir, Suryavarman's construction workers started to eradicate what Çatalhöyük archaeologist Ian Hodder calls "history within history." Directly beneath the West Baray floor lie the remains of a 3,000 year-old settlement.[12] We know this because the West Baray dried out in May 2004, and École française d'Extrême-Orient (EFEO) director Christophe Pottier took the opportunity to excavate there. Right below the surface, he and his team discovered telltale human burials, crushed pottery, and even bits of fabric and bronze that hinted at people living there in the early first millennium BCE. Suryavarman's 11th-century workers wouldn't have seen these specific remains, but it's very likely they uncovered other evidence that Suryavarman's reservoir was built on top of extremely ancient proto-cities. Building the West Baray meant unburying thousands of years of historical settlements, and then reburying them under millions of gallons of water.

Suryavarman had come from outside the royal family to take the throne, and he was also the Khmer Empire's first Buddhist king. Perhaps he wanted to signal that a new era had begun, and he did it by erasing history beneath his new artificial seas of creation.

Debates rage among archaeologists about whether the West Baray was useful or cosmetic. Obviously this water would have been important for drinking and farming, but we can't be sure how much of it actually reached people's homes and farms. The city was already riddled with other canals, and each city block had its own water collection pools that look like masses of tiny pinpricks on lidar maps. So it's not impossible that the West Baray was largely ceremonial. This interpretation would fit with evidence suggesting the West Baray was probably something of a boondoggle. To keep it perfectly in line with the east-west orientation established by the East Baray, it was built on a slope that kept the west end underwater while the east dried out. It rarely looked full. This pattern continues today, leaving the reservoir looking like a half-eaten rectangle whose waters almost never reach the ceremonial road to the king's palace.[13]

To get an idea of the size of Angkor's reservoirs during the city's heyday, Evans and I took a small boat into a medium-sized reservoir from an elevated walkway near the temple Angkor Thom. At one time, the now-empty East Baray would have been directly south of the waters where we floated, a light rain disturbing the lily pads nearby. Even though this reservoir was half the size of Yasovarman's gigantic East Baray, it was sizable enough that I felt like I was boating on a natural lake. Evans peered around, pointing out the fine construction of the reservoir's retaining walls. It brought us back to discussing the poor engineering at the West Baray.

I wondered aloud if there were engineers beating their heads against the wall a thousand years ago when their king told them the West Baray had to be oriented east-west.

"That would never be in an inscription," Evans laughed.

His offhand joke foregrounds one of the problems of studying urban life at Angkor. The roughly 1,200–1,400 temple inscriptions we have from this civilization only capture a tiny part of this city's story. We know the spiritual traditions of Hinduism and Buddhism influenced the arrangement of the city's east-west orientation, its layout following the track of heavenly bodies across the sky. But we don't have any writing left by engineers or construction workers telling us what they thought about building a reservoir that was obviously uneven. More poignantly, we don't know what *khñum* thought when they had to demolish their neighbors' homes to make way for a reservoir that might have been purely decorative. Based on the work they did, however, it's obvious that many people were willing to buy into the Angkorian system of debts and rewards. What made all that free labor worth it?

UCLA archaeologist Monica Smith has excavated a number of ancient cities, and believes that their lure is social. Like Stark, she argues that cities grew out of ritual centers that villagers traveled to once or twice in a lifetime to meet strangers and experience novelty. But as cities grew, people settled there because they wanted that kind of excitement year-round. It wasn't about spiritual pageantry anymore; it was about day-to-day interactions with thousands of other people. "Only cities could make that opportunity for intense interaction permanent and for a much greater range of purposes—social, economic, and political—than could ever have been envisioned for a ritual space," Smith explains.[14] Villages were enclaves of familiarity and sameness. But, she writes, "In urban settlements, unfamiliarity became the measure of human relations." Villagers who moved to work in Angkor were their own form of attraction for other would-be immigrants. Saskia Sassen, a sociologist who studies modern cities, echoes this sentiment, arguing that cities are places for delightful chance meetings and life-changing random encounters.[15]

It's worth considering that Suryavarman's mania for building city infrastructure may have served a purpose that he couldn't have understood. The more he induced *khñum* to expand Angkor's infrastructure, the more the city became a haven for the laboring classes. Santa Fe Institute network theorist Geoffrey West explores this idea in his book *Scale*,[16] based on his research into today's fast-growing cities. He's discovered that urban populations grow faster than their own infrastructure. West has found that doubling the size of, say, a city's water canals would more than double its population. Due to the benefits of sharing resources at high density, urbanites need about 15 percent less infrastructure than you'd expect based on population size. Put simply, urbanites multiply faster than their own urban spaces.

At Angkor, Suryavarman's focus on infrastructure would have

enabled an urban population explosion. The barays represented merely the most ostentatious parts of a canal infrastructure that redirected rivers from the Kulen Mountains, and Zhou reported that it allowed the Angkorites to reap the benefits of three to four harvests every year. The water infrastructure was in excellent shape, the farms were expanding, and so were the Khmer Empire's river-linked cities. If West is right about urban growth, we have to assume that Angkor's population was growing even faster than the city's footprint.

Wealth without money

What did all those urbanites do when they weren't farming or digging canals? The only contemporary description of life in ancient Angkor comes from Zhou Daguan's late 13th-century account, which he wrote in part as a travel guide to Khmer life for his Chinese audience. As a result, we learn little from him about what ordinary city dwellers thought, and a lot about how awkward it was for Zhou to use Angkorian toilets (no toilet paper!) and how hot the king's thousands of concubines were (he claims to have examined them pretty thoroughly from a balcony overlooking their quarters).

The Khmer themselves left behind over a thousand inscriptions on temple walls, giving us a tantalizing glimpse of how some Angkorians viewed their world. Unfortunately these writings are mostly temple bureaucrats praising their great leaders, or receipts of donations to the temples. But recently, data archaeology has given us a way to pluck the strands of people's everyday lives from these seemingly threadbare records.

Our ability to explore Khmer life through inscriptions in its people's native language is relatively recent. For over a century,

Westerners exploring Angkor focused on the more easily translated Sanskrit inscriptions, which consisted of poetic evocations of gods and praise for kings. Because Sanskrit was from Indic regions, these inscriptions became evidence for scholars who mistakenly believed that Khmer culture was "Indianized," basically a carbon copy of South Asian society. It wasn't until the Khmer linguist Saveros Pou translated the Old Khmer inscriptions that we gained full access to Angkorian history.

The Old Khmer language was unique to the region, and the only examples of it that we have are from Angkorian writing. Pou was fascinated by these ciphers growing up in Cambodia, and in the mid-20th century, she moved to France to study with the linguist George Cœdès. After living in Southeast Asia for several decades, Cœdès had translated most of the Angkorian Sanskrit inscriptions, and published an influential book that popularized the idea that Angkor had been "Indianized" or "Hindouisé." Because Pou was more rooted in modern-day Khmer culture, she struck out on a different path. In the 1960s and '70s, she focused on the idea of a specifically Khmer linguistic tradition. She immersed herself in the Khmer version of the Ramayana, and eventually compiled the only dictionary we have of Old Khmer. In the process, she had to invent a transliteration system for the ancient language, since Old Khmer has its own alphabet. Then she painstakingly translated words from Angkorian-era inscriptions into French and English for modern scholars. Without Pou's work, and her corrective to the "Indianization" hypothesis, we would still be struggling to understand how Angkorians organized their labor force.

The Old Khmer inscriptions contain a lot of nitty-gritty details about Angkorian life that the Sanskrit verses never touch. Written in prose, they tell us about economic life, as well as the occasionally boring details of who owes what to whom. It's revealing that temple workers wrote about elevated topics such as religion in Sanskrit, while they described everyday transactions in Old Khmer. Economic dealings were linguistically separate from the elite pursuits of kings and gurus. Taxes may have been paid to temples, but people didn't write about spiritual matters and financial ones in the same way.

In 1900, the French explorer Étienne Aymonier dismissed inscriptions about *khñum* as "those interminable lists of slave names." His attitude is reflected in a lot of 20th-century scholarship about Angkor, which focuses almost exclusively on elite life. But recently, University of Sydney archaeologist Eileen Lustig has used data archaeology techniques to study those "interminable lists" in depth, creating cross-referenced databases of every word from every inscription to look for intriguing patterns.[17] One of the first patterns to leap out at her was that temple servant names were 60 percent female. Given that we see gendered divisions of labor going all the way back to Çatalhöyük, she believes that farming and other temple work was handled by women. There's evidence to suggest Khmer women were in charge of farming outside temples as well. So when we imagine life in the Angkor Thom enclosure, for example, we have to assume it would have been dominated by women.

At Angkor, it appears that people didn't have work weeks—they had work fortnights. An inscription from Suryavarman's reign lists a group of temple workers, organized in two-week increments:

> Slaves to provide what is due: tai Kanso; another tai Kanso; tai Kaṃvṛk; tai Thkon; tai Kañcan; si Vṛddhipura—these for the fortnight of the waxing moon. [For] the fortnight of the waning moon: tai Kandhan; tai Kaṃbh; si Kaṃvit; tai Samākula; si Saṃʾap; si Kaṃvai.[18]

Each person, such as Kanso or Saṃap, has a title like *tai* (female slave/servant) or *si* (male commoner), and a work shift during either the two weeks when the moon is waxing, or the two when it is waning. Temples also organized festivals and rituals around the moon's phases. At one temple in the 11th century, we find these instructions on what kinds of offerings the temple workers should give to the gods:

> At changes in the moon's phase: two pāda of melted butter; two pāda of curdled milk; two pāda of honey; two 'var of fruit juice; at the saṅkrānta, one thlvaṅ of milled rice; at changes in the moon's phase, only one je of milled rice . . .[19]

Here you can see amounts are measured in things like "pāda" and "thlvaṅ," whose sizes probably varied over time. There doesn't seem to have been a standardized set of weights and measures throughout the Khmer Empire. In addition, scholars have debated when the holy day of saṅkrānta happened; it may have been fortnightly or annually, depending on the region.

Paychecks, such as they were, arrived fortnightly as well. Some inscriptions show temples gave out rice and other food to their *khñum* every two weeks. Even political cycles were measured by fortnight, with some inscriptions suggesting that heads of state expected major economic transactions to coincide with the lunar cycles, including tax payments and land grants. We have to imagine Angkor running on a schedule where the work fortnight was deeply connected to festival days and statecraft. Temple staffs often included astronomers who plotted the course of the moon, keeping track of work shifts and festival days. They also decided which days during the fortnight were most lucky, and that no doubt had some bearing on when people made major purchases or donations to their local temples.

The fortnightly pay schedule raises a question about how laborers survived. If *khñum* were fed rice based on two-week shifts each month, what did they do for food during the other two weeks? We know that higher-ranking temple workers were allowed to eat some of those tasty-sounding fortnightly offerings to the gods, and local elites also got leftovers on feast days when people poured gifts into the temple coffers. Possibly *khñum* could skim extra rice off these offerings, too. But more likely, they simply went back home and lived with their families during their weeks off. At Angkor, that would have been particularly easy because the walled enclosures around temples were residential neighborhoods.

We know this based on evidence written into the earth. Evans' lidar scans showed that the earthen mounds of house foundations can be found in neat grids around temples. Curious to know more, Alison Carter dug up one of these mounds inside Angkor Wat's walls in 2015. She discovered what appears to be the remains of a brick stove, complete with ceramic vessels for cooking.[20] Chemical analysis revealed remains of pomelo fruit rind, seeds from a relative of the ginger plant, and grains of rice. This is what archaeologists call "ground-truthing," and it's further confirmation that temples were at the center of neighborhoods full of people engaging in trade, farming, textile manufacture, and other domestic tasks. The people who lived there were paying their taxes in labor, at least part of the time. But they were also part of secular communities. In these temple neighborhoods, women farmed right alongside monks who composed Sanskrit poetry about their king.

These neighborhoods may not have been exactly like their counterparts beyond the temple walls, but they give us a sense of what life was like for people like Kanso and Kaṃvit, who reported for duty every other fortnight. As Stark warned, these places were what made the empire great, but they were also its vulnerability. Keeping people in line is a lot harder than keeping water contained in a baray.

There was another strong signal that emerged from Lustig's inscription data, though it might be more accurate to call it a lack of a signal. None of the economic records in Old Khmer mention money. At the same time, inscriptions show temples selling big-ticket items: roughly 75 percent of exchanges recorded are for large plots of land, 18 percent are for khñum, and 7 percent are for services related to marking land boundaries, sort of the Angkorian version of land value assessment. The few remaining items would have been temple supplies.[21] It's not as if Angkorians didn't know about money yet. They traded with other kingdoms that used coins, and had plenty of metal if they wanted to mint their own. There's also evidence that earlier Khmer cultures may have used money. Lustig identifies a couple of pre-Angkorian inscriptions[22] in which the scribes used specific units of silver as a way to place value on a rice field and an enslaved woman. The Khmer also had advanced math (including the revolutionary concept of zero), and sophisticated ways of measuring debts, interest rates, and exchanges.

After Jayavarman II founded Angkor, however, we never see goods or people valued in silver or any other unit of exchange. So what was the equivalent of Angkorian cash? Maybe, say some historians, there was a widely agreed-on list of valuable items that could be used instead of money.[23] In some early 12th-century exchanges, we see that a piece of land was sold for "2 gold rings, 1 silver bowl, various units of silver . . . 1 vessel, 2 water vases, 5 plates, 3 utensils, 1 candle holder, 20 cubits of fine fabric, 2 fast oxen, 2 lengths of new fabric 10 cubits long, and 3 goats." A khñum and her four children were sold for "60 garments." Generally we see worth measured with lists of items like these, a combination of animals, people, metals, and finely made household goods. Money may not have been needed, because each transaction could be cobbled together on a case-by-case basis from standard luxury items. What this suggests is that elite wealth was measured in land and the tools (including people) required to work it.

Everyday financial dealings in the *khñum* class were different. When Zhou visited Angkor in the late 13th century, he described how the city streets were lined with women selling food and other goods from blankets spread on the ground. Customers used coins from China and other places, as well as rice, grain, and fabrics as forms of money. It appears that inexpensive items might be had for cash, especially in these more informal marketplaces. Of course, there's always the chance that wealthy Angkorians *did* use money, and the people writing inscriptions thought monetary values were so obvious that they didn't need to record them. Another possibility, suggested by Zhou's observation that women conducted all trade in the Angkor marketplace, is that dealing with money was part of women's work and therefore not worth noting.

Taken together, the temple inscriptions and Zhou's observations reveal something profound about the Angkorian state. There doesn't seem to be a centralized form of control over economic exchanges. Local kingdoms could set their own variable prices for land and *khñum* at temples, while ordinary people got by on a combination of barter and foreign coins. The system seems unwieldy for those of us brought up in fully monetized societies, but it makes sense for a civilization where land and labor power were the most valuable items a person could own. Certainly Angkorian leaders loved their gold and silver—and probably they traded precious metals with foreigners who coveted the stuff—but they were not hoarders of money. Instead, they were hoarders of taxpayers who could be used to transform land. Most

Angkorian kings got rich on endless supplies of free labor from people like *tai Kaṃvṛk* and *tai Thkon*.

The fragility of stone

Nearly 200 years after Suryavarman's labor force broke ground on the West Baray, King Suryavarman II took the throne. Like his namesake, he did not inherit his position but instead fought for it in a bloody battle of succession. A prince from an outlying kingdom in what is now Thailand, he took to Angkorian life and left behind one of the city's most famous monuments: Angkor Wat, a mountainous temple compound that lies to the south of the two great barays. Suryavarman II was not an expansionist king, nor a particularly great warlord, but he's remembered because he did an especially good job of maintaining the canals and roads that connected Angkor to other parts of the empire and lands beyond. It probably didn't hurt that Suryavarman II also made sure to include a lot of very flattering pictures of himself in Angkor Wat's many dramatic reliefs. He's the first Angkorian king to depict himself in art, and it's hard to forget the image of him sitting in his palace on soft rugs, while a bunch of servants hold several multi-tiered parasols over his head.

Suryavarman II's glamorous self-representation was entertaining, but I wanted to know more about the people holding parasols. That's why I spent a quiet morning with Damian Evans visiting the place where Angkor Wat was made. The walled complex at Beng Mealea, located about 50 kilometers northeast of Angkor, was another one of Suryavarman II's many construction projects. Today it is rarely visited, and restoration efforts have just begun on its nested square galleries, libraries, and canals that flowed through the entire temple and the palace at its heart, forming a floor of gleaming water between ornate walkways. Long ago, it was flanked by a baray that was twice the size of the complex. But today the reservoir and deep moat around Beng Mealea are piled with the gargantuan stones that once formed its walls. Access to the palace is difficult, and Evans led me through the rear western gate, careful to stay on the path because we'd been warned about land mines in the jungle.

Around us the landscape was bumpy with regularly spaced mounds. "You can see there's an unnatural topography here," Evans remarked, referring again to the concept of anthropogenic geomorphology, or ways that humans shaped the land. We were seeing all that remained of the commoners' neighborhood of wooden houses that once surrounded Beng Mealea. When we reached the eroded walls of the complex, Evans paused. "We are in the middle of a dense downtown, surrounded by temple staff housing," he said simply. I imagined the towering trees vanishing around us, replaced by roads lined by tidy, thatch-roof houses on stilts, smoke coming off the stoves located underneath their elevated living quarters. Children shouting and farm animals grumbling.

Then we mounted sturdy wooden stairs that led into the temple complex over mossy, collapsing corridors. Arched ceilings sheltered a gallery whose long windows were fitted with stone balusters that acted as blinds. Each baluster had been lathed into a rippling, fluted shape that cast complex shadows. At the center of the complex, stone floors had been floated on columns so that water could run beneath them.

Evans joined me as we climbed past rocks that gushed from doorways, the wooden stairs creaking slightly beneath our feet. When we reached the top of an outer retaining wall, I looked down into a deep stone canal that was once part of Beng Mealea's outer moat. Today it appears to flow with blocks the size of elephants, as if time had somehow washed the structure into its own dried-out waterways. It was still early morning, and the air was cool under trees that grew from piles of stately rubble. All we could hear were the songs of birds and insects. As Evans unrolled a lidar map of the compound, I rebuilt the ruins in my mind. Thousands of *khñum* bustled past us, entering or leaving via long promenades that faced the four cardinal directions. They shuttled between their neighborhood farms beyond this moat and the elites who lounged at the center of Beng Mealea's many causeways. They were doing all the usual tasks of temple maintenance, tending farms, and pruning the gardens whose scented flowers were a key part of many fortnightly rituals. But this area was so densely populated because it was no ordinary provincial outpost.

Beng Mealea specialized in an industry that was of particular interest to Suryavarman II. It was strategically located in a spot at the nexus of two major roads and several waterways. One road connected to sandstone quarries in the Kulen Mountains to the north, and the other to an iron processing center at the Preah Khan temple complex in Kompong Svay to the west.[24] Both Beng Mealea and Angkor Wat are built from Kulen sandstone, which I could see piled around me now as if it had just arrived from the mountains. Here at Beng Mealea, *khñum* worked in shipping, receiving, and sometimes goods processing, too. As sandstone came into the complex along the canals, they cut it down into blocks and kept it for use at their own temple complex or sent it along to Angkor. When iron came from Preah Khan, they put it on barges for the long journey down the Khmer Empire's artificial waterways to Angkor. It's likely they did the same for rice and other goods heading into Angkor from the productive hinterlands. Suryavarman II, himself a child of one of these provincial cities, would have been keenly aware of the key role Beng Mealea played in his empire.

Some researchers believe that Beng Mealea was a beta version of Angkor Wat. It's the first place where engineers experimented with new kinds of archways and high walls that are ubiquitous at Angkor Wat. There are also some similarities in urban patterning. Evans showed me how the lidar reveals symmetrical city blocks around Beng Mealea that look extremely similar to those at Angkor. Here in the hinterlands, though, the neighborhoods show a greater variety of shapes and layouts. Still, Angkorians would no doubt have felt at home on the streets here, with their row houses separated by fish ponds. It was also thanks to lidar that we know what Beng Mealea's chief business was. After the technique revealed deep sandstone quarries in the Kulen Mountains, it was easy for researchers to connect the canal routes and see where all the sandstone to build Angkor Wat and Beng Mealea had come from.

There's still a lot we don't know. The lidar surveys revealed two previously unseen structures that nobody has been able to explain so far. The first is a complicated rectangular maze pattern dubbed the "coils," "spirals," or "geoglyphs." These were first spotted outside the moat at Angkor Wat during the 2012 survey, but the 2015 survey revealed similar coils outside the enclosures at Beng Mealea and Preah Khan. At first glance they appear to be waterworks, but Evans and his colleagues dismissed that idea because they are too shallow and are cut off from the city's general waterworks. The reigning hypothesis is that these rectilinear coils were specialized gardens for growing plants used in temple rituals. The often-flooded channels might have contained lotus, while Evans and his colleagues write that the raised areas could have supported "aromatics such as sandalwood trees."[25]

More mysterious are the so-called mound fields found near some of Angkor's largest reservoirs and canals. Unlike the residential mounds excavated by Carter and her colleagues, these mounds aren't packed with ceramics and food remains. They are just mounds, clearly the foundations for an elevated structure or structures. Their locations suggest that they may have been related to the city's waterworks, but of course correlation does not equal causation. The coils and mound fields are reminders of how much we still don't understand about how the ancient Khmer built their cities.

Suryavarman II and his predecessors were nothing without all those commoners and *khñum* cutting sandstone, smelting iron, harvesting rice, and shipping it back to the capital. When I visited Angkor Wat, it was hard to see the temple's ornate towers as the legendary Mount Meru, center of the cosmos, towering over the glistening waters of creation. Instead, I kept seeing piles of stone created in the quarries and workshops of thousands of unpaid laborers. I went inside its pale walls with a swarm of tourists, and paid to leave an incense offering for the spirit of the city, who inhabits a gold-draped Buddha sitting on an ancient pagoda in the temple complex. As I perused the famous reliefs showing Suryavarman II going to war with the Cham in today's Thailand, I looked mostly at the bodies of the men who bore his litter. The more infrastructure Angkor's *khñum* built, the more responsibility their king bore to maintain it. And as Stark warned, the patronage and debt systems were always on the verge of toppling.

CHAPTER 9

The Remains of Imperialism

Angkor has been abandoned so many times, in so many ways, that loss has become synonymous with the city's identity. But we can place at least some of the blame for its "lost city" reputation on the French explorer Henri Mouhot, who wrote a famous account of visiting Angkor Wat in 1860. His travel diaries, published posthumously, became a sensation and sparked a fascination for Cambodian culture in France. But it was a very specific kind of fascination. Shortly after Mouhot's journey there, the French claimed Cambodia as a protectorate. Stories of how a brave French naturalist "discovered" the riches of France's new colonial acquisition played well back home, especially because Mouhot implied that present-day Cambodians didn't appreciate their own treasures.[1] In fact, Mouhot suggested modern-day Cambodians were too savage to have made such a city, and that it must have been built by ancient Egyptians or Greeks. Only European scientists could possibly be trusted to study Angkor, given that the Cambodians themselves had allowed it to rot away in the jungle. It was archaeology as the white man's burden.

That's the feeling that steered conversations about Angkor in the West for the next century or so. Not only was this line of thinking factually incorrect, but it also erased the complicated history of Angkor's transformation from a massive capital city to a remote pilgrimage site occupied by Buddhist monks. It's important to understand that the city never stood empty, even after the royal family left in the early 15th century.[2] During the 16th century, when the city was supposedly "lost," the Cambodian king Ang Chan commissioned the completion of some reliefs at Angkor Wat. A few decades later, the city was described by a Portuguese friar named Antonio Da Magdalena, who was likely its first European visitor, roughly 300 years before Mouhot. In the 17th century, a Japanese pilgrim drew a map of Angkor Wat, and in the 18th century a Cambodian dignitary built a stupa for his family on the grounds of Angkor Wat. All these pieces of evidence suggest that people all over the world knew about Angkor, and it was a thriving pilgrimage destination. When the French colonizers arrived in the 19th century, they had to clear out a community of monks living on site. Mouhot's account was an act of revisionist history as audacious and long-lasting as Suryavarman's sweeping erasure of old Angkor beneath the West Baray.

The French craze for Southeast Asian civilization, inspired by Mouhot, reached a peak in 1878. In that year, the Paris World's Fair featured an exhibit of ancient Khmer art that French scholars had removed from Angkor and other Khmer Empire sites. In 1900, a group of French researchers made a permanent home in Southeast Asia by founding the École française d'Extrême-Orient in Hanoi. Then, in 1907, the EFEO took over supervision of archaeological work at Angkor, and their role continues to this day. In the early 20th century, this added academic credibility to the popular notion that the French had discovered Angkor and knew the most about it. It also led to French scholars mistakenly assuming that Angkor Wat was a European-style walled city, and that the Khmer lacked their own distinct cultural traditions.

A lot has changed since 1907. Damian Evans was affiliated with the EFEO when he led the lidar mapping project, which provided evidence to contradict European scholars' claims that Angkor wasn't as large as its inscriptions suggested. Meanwhile, many modern scholars have debunked the idea that Angkor had somehow gone missing, and it just so happened that a member of its colonizing nation "found" it in time for his colleagues to loot its temples for the world's fair.

As always, the truth is weirder and more complicated than the legend.

The first flood

Written into the ancient footprint of Angkor are dramatic signs of what went wrong in the city. Both lidar scans and excavations show frenetic—and increasingly complex—repairs and modifications to the city's canals, embankments, moats, and reservoirs over the centuries. As Evans pointed out when we visited the West Baray, some of these modifications were necessary because kings and their priests wanted the cityscape to line up with the idealized proportions of a cosmology. But some were responses to climate instability, as well as the usual wear and tear that infrastructure suffers when it's used by hundreds of thousands of people. Still, when the devastating floods finally came to Angkor in the 14th century, it might not have been obvious to people that their city would never fully recover. That's because Angkorians had been through much worse, and had come out of it with an even bigger empire than before.

In 928, mere decades after Yasovarman moved the capital city to its present location at Angkor, a king named Jayavarman IV took the throne. For reasons we don't fully understand, he uprooted the entire court and moved the capital again—this time, northwest to a bustling city called Koh Ker. There, the king ordered his *khñum* to build palaces and public works from enormous blocks of locally quarried sandstone. His crowning achievement, of course, would be a baray the likes of which nobody had ever seen. Workers erected a 7-km embankment across the Rongea River valley, blocking several large rivers, and creating what looked like a vast sea. Lined up perfectly to the south of the embankment, Jayavarman IV's workers built the tallest temple in the kingdom, Prasat Thom. An impressive pyramid, its stepped sides create the illusion that each level is receding into the distance, as if it were a tiered slope. At the top of Prasat Thom was an enormous linga, possibly made of bronze or wood, that disappeared long ago. Only the pyramid remains, each tier bursting with greenery, next to a river valley still strewn with the bricks blasted across the land over 1,000 years ago when the river overtopped the embankment twice within a few years, and then broke through completely.

It seems clear that Jayavarman IV wanted to create a very specific experience for travelers entering his city. Koh Ker was along a major highway between Angkor and Wat Phu in today's Laos. The embankment wasn't just a water retention feature; it was also a stunning promenade that led straight from the road to Angkor into the heart of Koh Ker's temple district. People traveling south along the baray to the capital city would soon see the giant linga that topped Prasat Thom directly in their line of sight. It would have loomed in the distance for the entire 7-km walk, a sign of fertility, sanctity, and power.[3] Needless to say, it's clear that Jayavarman IV's intentions were as much about political theater as they were practical. He wanted his dike to connect to existing roads, and he wanted it to line up with his temples. Like the West Baray, it was aspirational water management rather than solid engineering.

Why did Jayavarman IV put so much energy into building yet another city, when he could have kept expanding Angkor? One theory, put forth by French archaeologist George Cœdès, was that Jayavarman IV had usurped the throne and didn't care about Angkor. But Royal University of Phnom Penh historian Duong Keo argues that this once-popular idea was an example of Western scholars misunderstanding the rules of succession in Southeast Asian civilizations. Expecting that the "proper" order of things would be for a king's sons or brothers to take the throne after his death, historians failed to notice that this type of family succession was the exception rather than the rule at Angkor.[4] Duong suggests a more realistic possibility, which is simply that Jayavarman IV was from Koh Ker and had already built a fancy palace there. Why move? This interpretation also fits with recent environmental studies of the region, which show that people were farming and clearing the land with fire for centuries before and after it was the capital of the Khmer Empire.[5]

Assuming that Jayavarman IV wasn't an Angkor local, that would put him in the same category with many of the empire's expansionist kings, like Suryavarman I and II, who also hailed from distant regions. Usually these outsider kings were able to unite disparate regions in part because of personal connections to the provinces. At Koh Ker, we see evidence of a similar kind of allegiance-forming, but with a very different set of allies. We find those "interminable lists of slave names" at Koh Ker—there are thousands of them—and inscription expert Eileen Lustig believes they hint at where Jayavarman IV's sympathies lay. It was a period of great unrest and infighting in the region, and the Khmer Empire had shrunk dramatically. Aspara National Authority researcher Kunthea Chhom[6] describes how the lists of names at Koh Ker reveal a richly textured social structure in the city, including a wide variety of titles for the *khñum*.

Lustig believes it's possible that Jayavarman IV allied himself with a class of commoners called *si*, who are represented in Koh Ker inscriptions as being superior to commoners called *gho*. Jayavarman IV may have been elevating certain working-class people to serve as his allies. Suddenly those lists of slave names no longer seem "interminable" but instead a way to understand hierarchy among the working classes. The king's alliance with *si* may also have represented a larger shift in the Khmer social hierarchy, similar in some ways to the changing roles of *liberti* in early Imperial Rome. Lustig writes:

> The move of the centre from Angkor to Koh Ker might be best understood as a strategy by Jayavarman IV to weaken an opposing group and bolster his power base, allied with *si* [commoners]. Following the return from Koh Ker to Angkor, a change in the socio-political power structure, perhaps seen in the rising influence of many officials and *gho* commoners, becomes apparent.[7]

While his aristocratic neighbors squabbled among themselves, Jayavarman IV retreated to live among the commoners at Koh Ker. His next move may have been to ally with a subset of his own *khñum*, the si, to create a new kind of city, and to shore up his power. After his reign, when the capital returned to Angkor, a different group of khñum called the gho rose to power. Paying careful attention to slave names allowed Lustig to discern how elites allied themselves with different groups of laborers.

Still, the glory days of Koh Ker were short-lived. To figure out what happened to the empire's greatest baray, a group of archaeologists and civil engineers used their knowledge of present-day dam structures to re-create what went wrong with the embankment/road that marked the northern boundary of the city. Though there were myriad problems, it appears likely that the main issue was that engineers hadn't built an adequate spillway. During a rough monsoon season, the waters ran faster and higher than expected. In the end, the dike wasn't just overtopped: its damaged stone structure was eaten away by fast currents and torn apart.[8] There's evidence that people tried to shore up the failing spillway afterward, raising the wall's height for hundreds of meters around the breach, but repairs were never completed. Within a couple of years, the water overtopped the spillway again. Parts of the city were completely inundated, and it seems that the king gave up on his mega-baray. In 944, the capital returned to Angkor. A small number of people continued to live and farm in the diminished version of Koh Ker for centuries, connected by road to the capital at Angkor.

The story of Koh Ker is a microcosm for what happened eventually at Angkor itself. Beset by political woes from outside, facing a crumbling infrastructure inside, the city transformed from a dense hub to a rambling sprawl of farming villages. Still, it would be several centuries of expansion before Angkor went full Koh Ker. During that time, Suryavarman's laborers built up the empire's trade routes and bureaucracy, while Suryavarman II improved its infrastructure.

And then, in 1181, Angkor underwent its most profound urbanization. A new king rose, Jayavarman VII, who is still often called "the great king." His workers built thousands of roads, hospitals, and schools. His reign is referenced so frequently in Angkorian history that archaeologists refer to him by the nickname J7. He was Angkor's most successful expansionist. He was also an outsider who spent many years in exile among the Cham people in what is now Vietnam, before returning to Angkor during the Khmer-Cham war. Working with allies from within the Cham forces, he brokered peace, stopped a Cham invasion of Angkor, and took the throne. J7 ordered his temple scribes to create inscriptions about his commitment to peace, but he also conquered vast parts of Cham territory by force. He was a mass of contradictions, but J7 remade the Khmer Empire, and it's his city plan whose remains archaeologists study today. After his regime ended, we enter the final phases of Angkor's transformation.

The king of a thousand faces

Piphal Heng's fascination with Angkorian archaeology started when he was growing up in Cambodia. He told me that when he was a kid in the early 1990s, there were very few tourists at Angkor Thom, where King J7's famous Bayon temple resides. "When I was 11, my family went to a pagoda nearby," he recalled. "While they were doing rituals, I went up to the temple. By the time I reached the central tower, I got lost. I was frightened. There was nobody there, and I was surrounded by huge faces." I immediately understood what he meant. When I visited nearly three decades later, the Bayon was full of tourists, but the place was still magnificently disturbing.

One of Jayavarman VII's many architectural achievements, the Bayon has no walls. Its sprawling galleries are held up with a thick forest of pillars, topped with swollen, flower bud–shaped towers of different heights. From a distance, it looks like a jungle skyline. During J7's time, the place would have been painted in white and gold, gleaming like a pale lotus flower in the center of a manicured neighborhood for the hundreds of priests, artisans, servants, and family members in the king's retinue. Today, its algae-riddled sandstone[9] is the gray-brown of aging tree trunks. Wild trees have overtaken the gardens and pools where locals and pilgrims once strolled. But climbing to the upper terrace is an exercise in awe and dread. It's because of the faces.

When J7 took the throne, he became the Khmer Empire's second Buddhist king after Suryavarman. There was one key difference. Suryavarman had tolerated Hinduism among his subjects. J7 made Buddhism the official state religion.[10] Inscriptions suggest J7 declared himself Buddha incarnate, much the way his distant predecessor Jayavarman II had anointed himself a Hindu god-king in 802. During his reign, J7 ordered his court sculptors and engineers to fill the kingdom with Buddhas that many scholars believe wear the king's face. More likely, the intention was to merge the face of J7 with that of a bodhisattva, suggesting a perfect blend of state and religious power. Over 200 of these blended faces fill the Bayon, most stretching to the height of a full-grown person.

Pillar after pillar is built from four J7 faces, each pointing in a cardinal direction, projecting a sense of blissful repose. When I first saw them from a distance, they encouraged a sense of peaceful reflection. But as I picked my way up to the central tower, always reencountering that face, I began to feel like I was under surveillance. It was as if J7 wanted the Khmer to know they were under his watchful eye, subject to his judgment. By the time I reached the top of the shrine, it felt like every surface was a face. J7 hadn't merged with a bodhisattva. He'd merged with the city's infrastructure itself.

If Miriam Stark is right that people were drawn to Angkor for its pageantry, we have to assume that the Bayon was one of the places they most wanted to see. Millions of people were exposed to its message over the centuries. But that message wasn't just in the king's omnipresent faces. It was also in the pathways everyone took through the city to reach the Bayon. For distant aristocrats who were in J7's circle of patronage, he devised a clever plan to keep them returning to his domain. According to Stark, the Bayon held 439 niches for individual statues. "Scholars suspect these statues were *Jayabuddhamahānātha* images (statues of the Bodhisattva *Avalokiteśvara*), distributed by the king to at least 23 provincial centres named in inscriptions," she writes. "Their caretakers were required to bring the images to Angkor for annual consecration." It was the perfect excuse to require local leaders to report to J7's home turf, and it turned Angkor Thom into a site of holy pilgrimage. Angkorians witnessing the ritual would have understood that distant leaders traveled to the Bayon to pay homage the same way they did. Everyone was a servant to the king.

From the air, the Bayon looks like a stack of square blocks inside the bigger square wall of Angkor Thom, J7's sumptuous palace located directly between the West and East Barays. Ordinary pilgrims would likely pass through the eastern gate of the walled enclosure around Angkor Thom, following a perfectly straight road connecting the gate to the Bayon. They would stroll by a dizzying number of statues—nagas with their many snake heads forming fan shapes, proud garudas with their fierce eagle wings spread wide, rows of demons and gods—and in the distance they would glimpse the fancy gardens, pools, and homes of J7's household. Before entering the Angkor Thom enclosure, however, visitors would already have crossed part of the city itself. To the south was the far punier walled enclosure of Angkor Wat. North of the sprawling rectangles of the city's two biggest reservoirs, the east and west Barays, J7 added his own baray, the Jayatataka. Plus there was a wide canal running along the eastern wall of Angkor Thom that was actually a diverted branch of the Siem Reap River.

People coming to the city in J7's time would have also noticed that neighborhoods in the city center were as orderly as the upscale, walled temple neighborhoods. Like previous expansionist kings, J7 brought labor power to the capital. It appears these people altered the city's layout, creating a distinct tic-tac-toe-style grid that would be familiar to anyone who has hiked through Manhattan. Streets would have been lined with dense but orderly rows of wooden houses. This degree of coordination, Heng speculates, would have required a centralized urban planning authority. In one short inscription written by J7's son, we see a hint of what this meant. "It says Jayavarman VII took the land by force," Heng said simply. To transform the city's more organically shaped neighborhoods into a grid, the king's forces relocated his subjects. This kind of coercive urban planning was also a good way to suppress rebellion, which was a constant problem for the expansionist king. "One way to crush rebellion is to take people's property and make their families into servants of the temple," Heng mused.

To leave behind the urban footprint that he did, J7 seized property from people in the hinterlands, and marched laborers throughout his empire to build his famous public works. He may have been Angkor's greatest king, but history suggests that he may have toppled the house of cards that was the Khmer Empire's system of debt and patronage.

Climate apocalypse

When J7 died around 1218, his son Indravarman II took the throne briefly, and witnessed the first stages in Angkor's slow-motion transformation from urban center to rural pilgrimage site. During the next two centuries, Khmer holdings shrank dramatically; the empire lost territory to kingdoms in the places now called Laos, Vietnam, and Thailand. But Angkor remained at the heart of a stretch of territory that was undeniably Khmer, and Angkorians were close trading partners with neighboring groups as well as China and beyond. For people living in the city, life was still pretty good, especially if you were upper class. It's useful to recall that it was in the late 1200s that Zhou

Daguan wrote his famous account of Angkor, representing it as both prosperous and culturally vital.

But as the profound loss of Khmer territory attests, the patronage system was crumbling under its own weight. Khmer kings would have relied on a greatly diminished labor force, mostly taken from Angkor and its close environs. Still, it wouldn't have been hard for optimists living in the city to look around, see the crowded streets and ever-expanding canal system, and tell themselves that everything was going to be okay. To chart the city's transformation, we can look at its ever-morphing water infrastructure. During the 13th century, laborers made the canal system denser and more complex. They added more artificial waterways that stretched farther north, connecting the city to rivers flowing out of the Kulen Mountains, diverting their natural western paths to the south. Once this mountain water hit city limits, it was redirected into multiple canals and reservoirs, generally flowing downhill in a southeastern direction. But there was a problem.

Geologists have found that sediment from mountain runoff began to block key points in the canal network where rivers entered the city. That meant water was choked off before it could flow into the city's main canal system, and this problem led to more frantic canal-building. But then, as we hit the late 14th and early 15th centuries, there's a sudden, intense proliferation of channels going the other direction. These new canals dumped large amounts of floodwater *out* of the city's infrastructure and into the Tonle Sap.[11] University of Sydney geoscientist Dan Penny, who has explored the environmental factors leading to Angkor's demise, calls it a "cascading network failure."[12] Put simply, troubles at one crucial juncture in the network caused multiple catastrophic failures downstream.

The culprit behind this network failure was climate fluctuation. Penny writes that the late 14th and early 15th centuries presented Angkorites with incredible challenges. A multidecade drought led people to build many extra canals to siphon as much water as they could out of the mountains. But the drought abruptly ended with several years of unusually intense rainy seasons, which had two disastrous effects. First the rain overwhelmed a system designed to bring as much water as possible into the city, causing floods and the need to build those massive runoff canals into the Tonle Sap. Second, the monsoons rapidly eroded the dry, dusty landscape, sweeping tons of debris into the canal system. And that caused sediment to build up and block the water supply when it was needed. Adding to Angkor's troubles, the floods were followed by another multidecade drought.

There are a number of parallels to the modern world here. UC Berkeley public policy researcher Solomon Hsiang studies the economic effects of climate disasters, using examples that are both ancient and modern. When regions are hit repeatedly by storms, he told me, "no matter how wealthy a country is . . . they never quite make it back to baseline GDP." Repairing the infrastructure is so expensive that it's impossible to return to their previous economic baseline. And with each subsequent storm, their GDP is eroded even more. He calls this "sandcastle depreciation,"[13] and noted that any civilization, no matter what level it is at, will slowly melt away under this repeated onslaught. Angkor likely suffered from a version of sandcastle depreciation, with each hit to its infrastructure leaving the region less prosperous overall.

Hsiang's scenario allows us to imagine the slow apocalypse that overtook Angkor as a series of economic setbacks exacerbated by environmental crisis. The floods at Koh Ker were the first sign of what was to come, but the king avoided dealing with the fallout by moving the capital back to Angkor. Later rulers dealt with water shortages and silting by building more canals. Each time a king ordered laborers to dig another canal, we have to assume it was in response to farms being parched by drought or failing water infrastructure. At that point, we have to assume that Angkor's citizens might begin relocating to areas where farming was easier—and they didn't have to pay taxes in labor every year. Each time the city went through a climate crisis, the exodus of people was equivalent to a loss of money.

When the city flooded multiple times, there were enough *khñum* to dig new canals quickly, leaving behind a palimpsest of flood runoff infrastructure. But their frantic work wasn't enough; houses and farms were destroyed, and more people would have moved to less disaster-prone regions. Damian Evans, who tracked the city's ever-changing waterways via lidar, likens this stage in Angkor's history to what cities are dealing with right now. City planners are struggling with centuries of "legacy infrastructure" that wasn't built to withstand extreme conditions caused by climate crisis. "Archaeology provides this perspective where you see this is a recurring problem," Evans said. Sewers and waterways are hard to change—especially when they've been routed underneath roads and city blocks—and it's incredibly hard to adapt them to new environmental circumstances. At this point, there were fewer economic opportunities than ever in Angkor, and the city was no longer a beacon for people seeking the company of fascinating strangers.

As if Angkor's troubles weren't bad enough, armies were at the gates from Ayutthaya in today's Thailand. It was a good time to attack. The city was hemorrhaging labor, and its defenses were weak. Ayutthaya troops stormed the city in 1431 and occupied it for a few years, adding political instability to the city's roster of woes. Fed up with the lack of servants, endless floods, and demands from foreign soldiers, the Khmer royal family and court had had enough. In the mid-15th century, they moved the capital out of Angkor to a region near Phnom Penh. The Ayutthaya left, too.

Despite what many popular accounts will tell you, this is not when the city "fell." The upper classes had abandoned it, and taken their rules about debt slavery with them. Artists, priests, and dancers left for other cities, some in Ayutthaya. But the city's laboring classes stayed. Evans pointed out that people at Angkor in the 15th century repaired a major bridge by reusing stones from a 14th-century temple. The old Angkorian way was to quarry stone at Kulen, forcing servants to process it at Beng Mealea before sending it down the canals to Angkor. After the city's elite were gone, recycling was far more appealing to the people doing repairs. Yasovarman had once dismantled commoner neighborhoods to make way for the East Baray. Five hundred years later, commoners tore down the elites' monuments to repair the infrastructure their ancestors built.

To prove that point, Stark teamed up with Alison Carter, Piphal Heng, and several other researchers to publish a paper in 2019 about new discoveries in the Angkor Wat temple enclosure neighborhoods.[14] During excavation, they found the remains of households where people lived long after the royals had left the building. *Khñum* communities were still alive and well after the so-called fall of the city. Also in 2019, Dan Penny published a paper with Evans and two other researchers, geoscientist Tegan Hall and archaeologist Martin Polkinghorn. They synthesized two decades of evidence about Angkor's life cycle from lidar and ground-truthing. It has a title that sums up their findings nicely: "Geoarchaeological Evidence from Angkor, Cambodia, Reveals a Gradual Decline Rather than a Catastrophic 15th-Century Collapse."[15]

The one-two punch of these papers has changed the narrative about Angkor. There was no sudden tipping point; the city shrank slowly, and its people drained away over centuries. None of these researchers deny that there was a collapse at Angkor. It just had a very slow tempo. The cause was a trash fire fed on a toxic mix of bad leadership, bad city planning, and bad luck.

Piphal Heng is fascinated by the way Angkor's transition seems to mirror the Khmer people's transition from J7's Mahayana Buddhism to today's widely practiced Theravada Buddhism in Cambodia. "Today, Buddhism still revolves around the court, but there's only one Buddha. Kings are not Buddha," he told me. "It's a different mentality." What's also different is that Theravada Buddhist pagodas are owned by their local communities. This new way of practicing Buddhism broke the chain of inheritance that kept wealthy families and priests associated with temples for generations. Under Mahayana Buddhism, Heng explained, temples were handed down through elite families who used them to lay claim to land and slaves. But under Theravada Buddhism, "the monk's family ties are broken," Heng said. He can no longer pass along the temple to his family because "the temple belongs to the community and the community belongs to it." Heng believes that shift in beliefs played a major role in what changed at Angkor during the 13th and 14th centuries.

As Angkor's population left in what archaeologists call an urban diaspora, they returned to village life centered around Theravada Buddhist pagodas. There are parallels here to Çatalhöyük, whose people scattered from a dense urban core into small villages on the Konya Plain. Stark writes that the Lower Mekong Basin filled with a "rural agrarian system of hamlets and small towns whose farmers and artisans continued to pursue their livelihoods: perhaps with less direct state intervention." What collapsed wasn't Angkorian civilization, but "the political and urban core of an elite."[16]

Even after this transformation, there's evidence that the royals tried to move back to Angkor in the 16th century. Noel Hidalgo Tan is an archaeologist with the Southeast Asian Ministers of Education Organization Regional Centre for Archaeology and Fine Arts, and he made this discovery by accident while working as a student on a dig at

Angkor. An expert in ancient rock art, he left the dig one day to take a break and wander through the temple's upper levels. There, he found markings that looked to his trained eye an awful lot like faded artwork on stone. He snapped a few pictures and took them back to his lab. Using a specialized digital technique called correlation stretch analysis on the images, he was able to enhance the pigment colors. Suddenly, he was looking at pictures of elephants, orchestras, and people riding on horses in an area that looked like Angkor. There were abstract designs and a picture of a Buddhist stupa where a Hindu-style tower once stood. These paintings seem to belong to a specific phase of the temple's history in the 16th century CE, when it was converted from Hindu use to Theravada Buddhist.

"My working speculation is that the capital moved south after Angkor was supposedly abandoned. But then King Ang Chan went back to Angkor in the 16th century to reestablish it as a capital," Tan told me by phone from his office in Bangkok. "There seems to be a lot of other evidence that there was a flurry of activity in Angkor in the 16th century. You have inscriptions from that time saying the king turned Angkor into a Buddhist temple." He thinks the inscriptions, along with the picture of the Buddhist stupa, were a clear sign that these paintings referenced a Buddhist king's effort to revitalize Angkor. Apparently, the effort failed and Ang Chan returned to the capital at Phnom Penh. But this is further evidence that even as most of Angkor's population returned to village life, a number of them remained behind, too. The city lived on, but was increasingly a monument to its past glory.

One of the most moving and incredible monuments I saw in Siem Reap was neither temple nor palace. It was a nondescript complex of modernist Khmer buildings that serve as warehouses, many outdoors, for priceless statues from Angkor. Some are being restored, but most are here to protect them from looters. Some bear scars and score marks from where looters started to break them up before being caught by the authorities.

Thanks to Evans' connections to local archaeological authorities, I was able to go inside the warehouse to see its hundreds of Buddha heads, demon heads, and inscriptions. It quickly became clear that this place was the opposite of those forgotten storage places that hug the sides of US freeways. It was a living homage to Khmer history, almost a holy place. The Buddhas wore golden sashes; incense and candles burned at their feet. There was a shrine to one particular Buddha, centuries old. Evans said it had survived an attack by the Khmer Rouge, Pol Pot's army, who were in the habit of blowing up Buddhas. The story goes that they strapped a bunch of land mines to this particular Buddha, but it survived intact. Only the seven-headed naga that once formed a protective parasol over the Buddha's head was blasted away. The workers at the warehouse restored the naga, and put the Buddha into a pagoda where incense and candles burn, and lotus offerings grace his feet.

Evans and I visited a similar kind of shrine at Preah Vihear, a massive cliffside shrine built under the orders of Suryavarman I in the tenth century. A millennium later, it was the last stronghold of Khmer Rouge forces, who surrendered there in 1998. After ascending through five temples, each higher on the cliff and more ornate than the last, I arrived at an escarpment overlooking the fields of Cambodia and Thailand. In a grassy field behind the fifth temple, there are the remains of bunkers, weapons caches, and the mount for a massive gun. The gun mount, almost the shape of a stupa, has been converted into a shrine. It was piled with flowers, metallic ribbons, burning incense, and other offerings. Preah Vihear is built in an area that's still contested, and Cambodian soldiers lounge everywhere, sometimes kindly helping the older Khmer visitors to step up or down the massive stairs between temple levels. A guard watched a YouTube video on his phone as we snapped pictures of ancient engravings. I stood at the intersection of recent events and deep history, wondering whether every city is doomed to churn endlessly through cycles of violent expansion and abandonment.

This question was on my mind when I returned to Phnom Penh, where the Khmer Rouge engineered a mass urban diaspora in the mid-1970s. It was hard to imagine such a vital city emptied of its inhabitants. Today, Phnom Penh's streets are thronged with vehicles, from huge SUVs and buzzing scooters, to tourist-crammed tuk tuks and cyclos. Every square centimeter of space was being used on the sidewalks, where people set up impromptu grills for making lunch next to piles of bricks and coal. Vendors pushing carts were selling everything from fruit and bread to toilet paper and coffee. There were no empty buildings, except in the expensive high-rises visible in the distance across the Tonle Sap River. Old movie theaters had been converted into shantytown warrens; a former French department store was crammed with apartments, and freshly washed laundry hung from lines inside its graceful roof displays. When people returned to the city, they built homes everywhere in the ruins, including beneath the roofs of massive old cathedrals and Buddhist wats. Everywhere I looked, the walls and streets and alleys were bursting with activity.

But as the memorials to victims of the Khmer Rouge attest, the city was violently purged of all its citizens just a few decades ago. Urbanites were sent to work in what came to be called the killing fields, forced labor camps that also served as mass graves. High schools and temples became torture and detainment centers. It made me think of J7, who remade the capital city, took people's land, and sent thousands of workers all over his empire to do his bidding.

Political disasters leave their marks on the land as surely as natural ones do. But over time, those marks become a palimpsest of testimonials to the ways people survived. The Khmer continued to live at Angkor long after their kings were gone, remolding the land until it resembled the farms and villages that had occupied it in the 700s. Likewise, the Khmer returned to Phnom Penh to reoccupy the city in new ways after Pol Pot's troops fled north to Preah Vihear. It's tempting to call this a cycle of repeated forgetting and repeating a dark history. But that's too simplistic. Another possible interpretation is that the Khmer urban tradition is more powerful than the forces that tore it apart. Angkor isn't a lost civilization; it's the living legacy of ordinary people who refused to give up.

PART FOUR

Cahokia

THE PLAZA

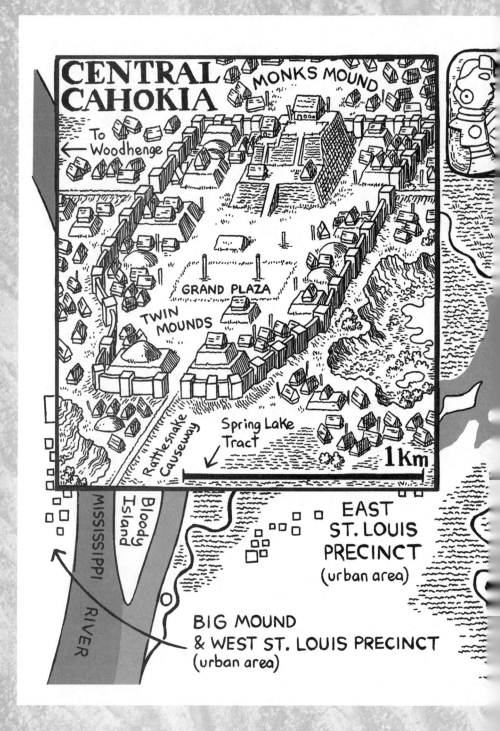

CHAPTER 10

America's Ancient Pyramids

A thousand years ago, huge pyramids and earthen mounds stood where East St. Louis sprawls today in southern Illinois. Majestic urban architecture towered over the sticky mud of the Mississippi River floodplains, and elevated walkways wound between densely packed neighborhoods, public plazas, and outlying farms. Ceremonial poles, painted and adorned with ritual objects, were planted in mound tops like signposts. The city was so impressive that word about it spread up and down the Mississippi and its tributaries, from Wisconsin down to Louisiana. Thousands of people came to the city, drawn by tales of its elaborate parties, pageants, and games. Some came to have fun, but others were in search of a new kind of civilization. Many visitors were so impressed that they never left.

The city became an immigrant sanctuary, its neighborhoods bursting with people drawn from cultures all across the southern United States. At the city's apex in 1050, the population exploded to as many as 30,000 people. It was the largest pre-Columbian city in what later became known as North America, and bigger than Paris at the time.

One particularly magnificent structure, an earthen pyramid known today as Monks Mound, marked the center of downtown. It towered 30 meters over the city. Cut into its southern slope were three dramatic ascending levels, each covered in ceremonial buildings. The entire mound occupied an area roughly the size of Egypt's Great Pyramid at Giza. Standing on the top level, an orator could be heard all the way across the 50-acre Grand Plaza at its southern base. A kilometer-long ceremonial causeway stretched southward from Monks Mound. This elevated path cut through flooded ground and terminated at another massive earthen structure, which archaeologists call Rattlesnake Mound.

Flanking the pyramid to the west was a circle of tall wooden poles, dubbed Woodhenge, that marked the solstices. To the east was one of the city's many deep pools called borrow pits, where Cahokians had dug up earth for their mounds. Most borrow pits were lined with colorful clay and designed to hold water seasonally. Aligned on a north-south grid, these artificial mountains and ponds, walkways and time-keeping poles, suggested to visitors that the city was inhabited not just by people, but by otherworldly presences who came from realms above and below the human world.[1]

Though the city's most visible monuments were its mounds— and there were hundreds of them, large and small, scattered over kilometers—its heart was the plaza. In the Grand Plaza at the foot of Monks Mound, city dwellers leveled the ground and covered it in a thin layer of gravel to accommodate crowds who came for ceremonies and sporting events. The place was roughly the area of 38 American football fields, and it was the template for many smaller public areas throughout the city. Some of the smaller plazas were little more than courtyards surrounded by a dozen homes in a neighborhood. Others would have rivaled the Grand Plaza. Urbanites kept all these plazas open and clean, ready for many kinds of social events. Plazas were a crucial part of the city plan because this was a community built on forming a special kind of public sphere—one where ideas could change the shape of the land, and vice versa. All cities offer their residents a chance to experience public identity—Çatalhöyük had its history houses, Pompeii its streets, and Angkor its temple complexes. But at Cahokia we see purpose-built structures throughout the city devoted entirely to the masses. Plazas were places for people to form crowds, whether to watch sports or listen to a sermon. They defined Cahokian society in the same way street shopping defined Pompeii. This was a city devoted to the transformative power of public life, full of carefully crafted meeting places where individuals could come together and become something greater than themselves.

Despite its grandeur, this city's original name has been lost to time. Its culture is known as "Mississippian" because its remains are found all along the great river that unites the southern and northern parts of the continent. When Europeans explored Illinois in the 17th century, the city had been abandoned for hundreds of years. At that time, the region was inhabited by the Cahokia, a tribe from the Illinois Confederation. Europeans decided to name the ancient city after them, despite the fact that the Cahokia themselves did not claim to have built it. The name stuck.

Centuries later, Cahokia's meteoric rise and fall remain a mystery. By 1400 its population had largely dispersed, leaving behind scattered villages on a landscape completely geoengineered by human hands. Mississippian culture can be found in the traditions of Siouan tribes, especially the Osage, and the mounds continue to inspire modern indigenous people from many tribes. But what led to the city's founding, and its abandonment, remain ambiguous. Looking for clues, archaeologists dig through the thick, wet, stubborn clay that Cahokians once used to construct their mounds. Buried beneath a meter of earth are layers of millennia-old building foundations, trash pits, cryptic remains of public rituals, and graves. Together, they tell the story of a civilization that may have been designed from the start to be temporary. For Cahokians, abandonment was not a failure or loss, but instead part of the expected urban life cycle.

Joining the movement

By the reckoning of the Roman calendar, people started erecting Cahokia's first monuments in the late 900s. At the time, European civilization was mired in the superstitions and brutal monarchies of the Middle Ages. But in North America, there was no entrenched medieval aristocracy, nor ancient Latin texts hinting at a lost great civilization. Instead, there were powerful but ever-changing social movements that temporarily united tribes and nations, and whose closest modern analogues might be political revolutions or religious revivals. And these unfolded against a backdrop of living urban history in the Americas, embodied in massive earthworks and stone monuments, whose origins went back thousands of years.

Based on what we know from indigenous oral histories and observations by Europeans in the 18th and 19th centuries,[2] it's likely that Cahokia was founded by leaders—or maybe one charismatic leader—who promised a spiritual and cultural rebirth. Some call Cahokia a city built on religion, but its origins were more complicated than that. Perhaps the best way to put it would be to say the city was spawned by a social movement that swept across the US south and Midwest, along the shores of the Mississippi River.

The Cahokians left no writing behind, so we can't say for sure what this movement was. But it was inspired by the founders' knowledge of

North American history. Mound cities are an ancient tradition in this part of the continent, going back millennia before Cahokia. North America's first known earthworks are in Louisiana. The oldest, called Watson Brake,[3] dates back 5,500 years—centuries before the first Egyptian pyramids were built. Another is at Poverty Point, built 3,400 years ago near the Mississippi in northern Louisiana. Today you can still see Poverty Point's crescent-shaped mounds towering like huge nested parentheses on a bluff overlooking a now-dry riverbed. A thousand years after Poverty Point was abandoned, people from the Hopewell culture built even more astounding mound cities in Ohio and throughout the northeast. The Cahokians would have known about these mounds from ancestral histories—and could have seen them along the Mississippi—but they might also have been influenced by contemporary pyramids in the Mayan and Toltec metropolises farther south.

The builders of Cahokia probably intended to build a city in the image of these previous civilizations. They also built it extremely fast, as if spurred on by enthusiastic belief. University of Illinois at Urbana-Champaign archaeologist Tim Pauketat has studied Cahokia for most of his career. He says that its mounds appear so abruptly in the archaeological record that it's as if they were built directly on top of a constellation of small towns that belonged to people known today as the Eastern Woodlands tribes.[4] As the city grew, so too did its farms, and the cultivated fields spread outward from Cahokia into the Illinois uplands. We find traces of Mississippian culture all along the river, where towns and small cities built mounds and shared some of the rituals of Cahokia. It's likely the city was something like Angkor, whose architectural styles and bureaucratic influence at some points reached thousands of kilometers beyond the city itself.

Cahokia was like Angkor in other ways, too. It had the urban design of a tropical city, with big stretches of farmland between neigh-

borhoods, and earthen mounds that became city centers. Early residents of Cahokia spread to both sides of the Mississippi, reshaping
the land with crops and earthworks. The city footprint was enormous,
and archaeologists sometimes say the metropolis had "precincts": the
densely populated center around Monks Mound, as well as another
center identified in East St. Louis, yet another where the city of St.
Louis stands today. It's likely these weren't separate cities; they were
more like downtown neighborhoods separated by farms.

Cahokia was built entirely with human labor. Workers used stone
tools to quarry clay in deep trenches that later became borrow pits, and
carried woven baskets of it to the growing bulk of the mounds. When
the clay was dumped out, they packed it down until the mounds were
as solid and substantial as mountains. Centuries later, archaeologists
digging into the sides of Monks Mound could still discern circular
clumps of clay, each a slightly different color, marking the spots where
basket loads were emptied.[5] Cahokians' backbreaking work on these
monuments may have been ritualistic. Perhaps they dug and carried
merely to enhance the city's greatness and power. Or perhaps they
were debt slaves, like the *khñum* of Angkor.

Unlike Pompeii, Cahokia didn't have streets lined with shops.
What archaeologists know of its urban plan includes no permanent
marketplace, nor merchant halls. And yet early 20th-century anthropologists had a hard time believing such a major city wasn't centered
around commerce or mercantilism. Partly they were inspired by Gordon V. Childe, inventor of the "Neolithic Revolution," who believed
that cities by definition had to have money, a taxation system, and
long-distance trade.[6] And, like early European explorers at Angkor,
they also assumed that every ancient city in the world was built with a
central marketplace and walls around it. But in the past few decades,
archaeologists like Pauketat have argued that the city was a spiritual

center rather than a trade center. As evidence, he points to the kinds of objects that people took home with them from Cahokia.

One of the most common items that people took away from the city was a distinctive form of ceremonial pottery, called Ramey, that was made exclusively at Cahokia. Ramey pots were aesthetically beautiful and technically complex. The clay was tempered with ground mussel shells, which kept its perfectly thin walls from developing cracks during firing. Incised with complicated designs representing the underworld, some Ramey pots also have delicate animal heads for handles and are painted in vivid, abstract spirals of red and white. They're found throughout Mississippian settlements, and are further evidence that people brought back symbolic items from Cahokia, rather than functional items like amphorae of wine or specialized tools.

Archaeologists have discovered other small souvenirs from Cahokia—figurines, decorative projectile points, and ceremonial beakers—as far away as Wisconsin and Louisiana. These findings suggest that Cahokia traded in ideas and spiritual tenets rather than practical commodities like food, tools, or textiles. No doubt people bartered with each other on a small scale, but this wasn't a culture built around commerce like at Pompeii. Cahokians came together to participate in a cultural worldview, and they bonded over a shared sense of public purpose. We can partly reconstruct that purpose by paying attention to the layout of the city.

Mississippian public life

Though it was massive, Cahokia's Grand Plaza was kept mostly empty, as if part of its function was to suggest all the ways that people might be joined together. Wooden screens and ceremonial poles could be set up for various activities, but there were no permanent structures like shops or temples.[7]

Some days, the Grand Plaza was cleared so that people could play a game with pucks and spears called Chunkey. Pauketat describes what he thinks it would have been like:

> The chief standing at the summit of the black, packed-earth pyramid raises his arms. In the grand plaza below, a deafening shout erupts from 1,000 gathered souls. Then the crowd divides in two, and both groups run across the plaza, shrieking wildly. Hundreds of spears fly through the air toward a small rolling stone disk. Throngs of cheering spectators gather along the sidelines and root for the two teams.[8]

Cahokian artisans made figurines of popular Chunkey players, and one shows a man kneeling to roll a puck, his hair drawn back into an elaborate bun and his earlobes stretched with decorative plugs. Based on figurines like this, and accounts from Europeans witnessing versions of Chunkey played elsewhere,[9] we know the game was as much about gambling as it was about athletic prowess. Chunkey players would roll their pucks into an arena, simultaneously throwing their spears. The winner was the player whose spear hit closest to where the puck came to rest.

But perhaps the real winners were all the people who bet on that player, and took home whatever prizes were on offer. Apparently the game was rather slow and involved a lot of gambling and audience participation. That made it a perfect sport for bringing people together who wanted an excuse to socialize. The game was so beloved that even the Chunkey pucks themselves became art, and people who journeyed to Cahokia often returned to their villages with one of the city's finely shaped and polished pucks.

Cahokia was also a city that loved to party. Festivals at Cahokia were mostly centered around group meals where people feasted on barbeques of deer, bison, squirrels, and even swans. Centuries later, archaeologists found huge party trash pits full of fire-cracked bones and broken dishes. Revelers passed around beautiful festival dishes full of fruits and breads, and used special ceremonial beakers to down a few swallows of "Black Drink," a caffeinated hallucinogen used during ceremonies to induce visions and vomiting.

It's likely the city's population doubled during these festivals because people came to Cahokia from cities all along the Mississippi. One sure indicator of this is that the Black Drink is made from holly trees that grow hundreds of miles away from Cahokia, so people had to bring it with them. Visitors brought their other valuables from home to share, too. Tools and pottery in non-Cahokian styles found their way into Cahokian trash heaps and sacrificial fires. When I visited the Illinois Archaeological Survey offices, I saw a projectile point that was carved in a southern style from the Texas area, but made from locally quarried Cahokia chert. This suggests an immigrant making weapons with hometown methods, using the same stone that Cahokians preferred for their projectile points. It was the stone tool equivalent of a modern-day Korean taco, whose delightful existence is thanks entirely to a history of cultural mixing.

But the revelry in the city wasn't always about sports, barbeque, and projectile-point style fusion. Giant festivals can also whip people into a frenzy of ecstatic belief in shamans or politicians or both. Public figures stood at the top of Monks Mound and addressed crowds in the plaza.[10] And then there were the shows. A cross between theater and ritual, these spectacles focused on stories of fertility and renewal, as well as tales of heroes and gods.[11] We can't be sure if people attending them were experiencing something that medieval Europeans would have called church, or something that contemporary Americans would call a Star Wars movie. Most likely, it was a bit of both, depending on the circumstances.

Cahokians built large earthen platforms like stages, where people dressed as mythic figures acted out stories to mark important times of year like the harvest. Some of these pageants included human sacrifices. These sacrifices could take many forms—I'll discuss them in detail later—but human life was not the only offering that Cahokians made to the gods during these performances. Archaeologists have found the bodies of sacrificial victims surrounded by many offerings, including the disinterred bones of ancestors that people brought to rebury with the newly dead. What happened next is reminiscent of the Death Pit at Domuztepe in Turkey. After the stage was heaped with bodies, bones, and riches, it was covered over with earth, and tamped down to form a peaked top like the one on Rattlesnake Mound. These peak-top mounds resembled the deeply peaked roofs on typical Cahokian homes. Often, people erected these stage/mounds at the edges of Cahokia's downtown plaza area, and some archaeologists speculate that they served as a special kind of boundary marker between our world and the world of the dead.[12]

Human sacrifices were no more out of the ordinary to Cahokians than the grisly executions of infidels were to their contemporaries in Europe. In both Europe and the Americas at this time, sacrifice was a public spectacle, used to solidify social norms and hierarchies. In European countries, executions in the town square were a way for rulers to show their power and purge their enemies. Centuries after the Cahokians stopped engaging in human sacrifice, England's King Henry VIII was famous for publicly executing his advisers as well as two of his wives. Early European settlers in the Americas also lovingly recorded their public executions of infidels at colonies in Plymouth and Salem. Like these European executions, human sacrifices at Cahokia may have served to reinforce a social hierarchy whose rulers stood on the top of Monks Mound.

Cahokians designed their city to reflect a fascination with astronomy. People at Cahokia tracked the movements of the stars, moon, and sun, often orienting their homes to the positions of these cosmic bodies in the sky. During the city's biggest population expansion, its street grid was aligned to be exactly 5 degrees off the north-south axis. Pauketat and his colleagues believe it is oriented toward an astronomical phenomenon called the lunar standstill,[13] during which the moon's altitude in the night sky rises and falls dramatically during a two-week period.

The city's boom years may have been jump-started by an even more astounding astronomical event. In 1054, just as the city was growing, a supernova lit up the sky for almost a month. It was so bright that it would have been visible during the day and as luminous as the full moon at night. We have records of this event across the world, from scrolls authored in China, to paintings on the walls of Chaco Canyon in New Mexico, where another indigenous urban civilization was booming. Pauketat believes it's possible that an enterprising group of religious or political leaders took the supernova as a sign that it was time to spread word about their burgeoning civilization. Perhaps the exploding star lent credence to a new set of beliefs that united previously disparate groups in a common purpose, laying the groundwork for what became Mississippian culture.

Whatever Cahokia's leaders did, they were incredibly successful at attracting a broad audience. Over a third of Cahokia's population were immigrants who had been born and raised far from the city.[14] We know this because scientists use a process called stable isotope analysis that reveals where a person grew up. By studying the chemical composition of tooth enamel from human remains at Cahokia, scientists can discern the specific isotopic signatures left behind by the food and water that people ingested as children. The process is often used in forensics, where it can help detectives figure out where a body has come from. In the hands of archaeologists, it reveals immigration patterns. If a person was buried at Cahokia, but grew up consuming food and water from a distant location, then that person was almost certainly an immigrant.

Cahokia may have drawn people in with its political power, but the city was also a place where humans did extremely mundane things, like farm, hunt, maintain infrastructure, and raise families. When archaeologists excavate here, they mostly find objects that come from those kinds of human activities: broken hoes tossed aside, gnawed deer bones from dinner, broken clay pots, and the telltale deep post holes that mark the edges of somebody's old wooden house. Still, the Cahokians created these everyday objects on a scale that was extraordinary for North America at the time. The city's farmlands, which produced several kinds of fatty seed grains, as well as fruits, squash, beans, and corn, fed more than 30,000 people at the city's height between 1050 and 1250. It would have been possible to walk roughly 19 kilometers from Monks Mound to the Mississippi River, take a canoe across, and continue walking for another several kilometers, without ever really leaving the city and its farms.

The lost crops of North America

Cahokia lies in a crazy quilt of ecosystems along the Mississippi River called the American Bottom. Rain and floods fill the area with seasonal ponds and swamps, while the surrounding bluffs give way to prairies perfect for growing food staples like maize and other starchy seeds. It is one of the most fertile stretches of land in North America, and Cahokians were well aware that they lived in a place so fecund it was practically magical.

One of the most intriguing clay figurines we have from Cahokia is known as the Birger figurine, and it was discovered in an eastern out-lying farming precinct called the BBB Motor Site (so called because it was uncovered during freeway construction). Found alongside other ritual materials, the Birger figurine is a deep reddish-brown flint carving of a woman farming on her knees, teeth gritted with effort as she wields a stone-and-wood hoe. But she does not till the land. Instead, her instrument cuts into the back of a serpent, its fat body coiled around her bent legs. In one powerful hand, the woman holds down the serpent's head, which resembles a snarling bobcat. Behind her back, its tail branches into vines heavy with gourds. Clearly, she has already harvested some of the serpent's bounty; strapped to the woman's back is a woven basket full of squash.

Washington University anthropologist Gayle Fritz describes showing the Birger figurine to traditional Hidatsa farmer Amy Mos-sett, who immediately recognized it as "Grandmother," or "the old woman who never dies," a powerful spirit who oversees the harvest.[15] Here we see continuity between Mississippian beliefs and modern-day Siouan ones, as well as a sign that farming wasn't just a job for Mississip-pians. It was a dangerous struggle to harness the power of otherworldly forces, as dramatic as any hunt or battle. At Cahokia, agriculture was part of an ongoing drama that involved life, death, and the cosmos.

Unlike indigenous farmers to the south, the people of Cahokia didn't plant maize until later in the city's development. Instead, they ate domesticated North American plants like goosefoot, little barley, marshelder, maygrass, and erect knotweed (not to be confused with its invasive cousin, Asian knotweed). These plants are sometimes called "lost crops" because they were once farmed intensively but have gone wild again. Cornell University archaeobotanist Natalie Mueller has spent several summers on a quest to track down the elusive remains of some of these lost crops, especially the one called erect knotweed.[16] Growing along rivers in the United States now, it looks like an unremarkable, stalky plant, with bright, spoon-shaped leaves. But when Cahokia was founded, knotweed had undergone a thousand years of cultivation by indigenous people across the south. Generations of farmers had selected for knotweed whose seeds were large, thin-skinned, and grew quickly—much the way corn grows faster and bigger after millennia of selective cultivation. Mueller has found centuries-old caches of these plump cultivated seeds buried in areas where people lived in the Mississippian region.

Domestic knotweed produced starchy, extremely hard seeds that had tough shells. To eat them, Mueller thinks, Cahokians cooked the seeds like popcorn in the embers of a fire, their tasty goodness popping out of the shells as they heated. Cahokians also could have used an ancient process called nixtamalization, in which they soaked erect knotweed in lime—the chemical, not the fruit—to turn it into a hominy-style porridge. Many indigenous groups in the Americas used nixtamalization to soften the hulls of maize before cooking, and it's likely Cahokians would have known about the technique. When they weren't enjoying popped knotweed, Cahokians might have eaten a thick, rich knotweed mush flecked with meat and spices. Knotweed and other lost crops formed the basis of a varied diet that combined fish and game with breads, porridges, oils, roasted nuts, stews, baked squash, and beans.

Fritz, who previously described the meaning of the Grandmother figurine, has spent most of her career studying the cuisines of indigenous people in the American Bottom where Cahokia emerged. She recalls learning about Cahokian life by exploring a giant garbage pit next to Monks Mound.[17] Digging through layers of debris, archaeologists found multiple layers of party goods, including swan bones and other barbequed animals, many kinds of seeds, shattered pottery, and even a layer of ants that likely came to feast before the refuse was covered with grass and burned. Fritz explains these are the remains of feasts from early in the city's life, when Monks Mound was new. The wide range of food items, many tossed out when they were only half eaten, gives clues about how Cahokia fed its people. A lot of the food came from farms many kilometers away, at places like the BBB Motor Site, where Fritz and her colleagues identified evidence for intensive farming alongside small settlements full of Cahokian ritual items.

The parties that filled the garbage pit, argues Fritz, also provide important hints about how city dwellers organized agricultural work and divvied up seasonal bounty. To reconstruct this complex social system, Fritz looks to plant remains as well as records from Europeans like Le Page du Pratz, who described monthly feasts among the Natchez, a large community of mound-building farmers in Mississippi in the 1700s. Both sources reveal a pattern where farmers brought harvests from the hinterlands into city centers for distribution during feasts. The question is: How was this distribution managed? Fritz argues that the Mississippians likely controlled land via kinship networks, the way the Hidatsa did, with many families sharing the same field. "American schoolchildren are taught that private, individual ownership of land was a concept foreign to Native Americans," she writes. "Nevertheless, it is clear that families or extended kin groups held exclusive use rights to firmly demarcated plots of land for farming."[18] Women worked the farms, planting small batches of crops in different parts of their plots, while men tended small gardens of tobacco grown next to homes within the city.

The fact that these distant farms fed the city dwellers raises another question. Were people at the BBB Motor Site offering tribute or a food tax to elites who lived on top of Monks Mound? I decided to pose this question to some archaeologists in the best place to talk about Cahokia's history: a pub in Edwardsville, Illinois, called The Stagger Inn. Founded by an archaeologist, it's known to Cahokia researchers simply as "the archaeologist's bar." Every Thursday, people working the digs all over Cahokia converge on the place for beer, hamburgers, and delicious fries.

At a battle-scarred wooden table next to a stage where musicians were setting up, I was joined by Tim Pauketat and Indiana University, Bloomington, anthropologist Susan Alt. I immediately started asking them about Cahokia's economic structure because I was curious about whether the city's elites persuaded people in the outlying farms to bring them food. Was there some kind of trade network? Pauketat actually rolled his eyes when I asked that. He and Alt were both very opposed to the idea that Cahokia might have been a trade center, and called it a mistake to view the city as an economic entity. "The primary purpose of the city was not trade or work. It was spiritual," Pauketat said, after I plied him with more beer. "Wealth isn't really the right word for what they had, but it was a side-effect."

Alt had further evidence that Cahokia was a place devoted to spirituality. She was excavating at a site devoted to spiritual rituals called Emerald. Located in St. Clair County, Illinois, Emerald might even have been the birthplace of Cahokian spirituality—it's full of

Mississippian artifacts but predates Cahokia's population explosion. "Maybe people came there, then immigrated to Cahokia and stayed?" Alt mused. If true, that would provide more evidence for the idea that Cahokia's founding grew out of emerging belief systems rather than trade concerns.

But there had to have been *some* economic system, I argued. After all, some people were growing food and other people were eating it. Was there trade with other cities along the river, or a marketplace where toolmakers from the downtown area could trade for maize from the uplands? Pauketat shrugged. "Sure, some people were specialized, or getting food from other people, but practices were heterogeneous. It would have worked differently in different neighborhoods." Maybe people in one neighborhood traded their Ramey pottery with another neighborhood that produced particularly excellent reed mats, he suggested. Maybe families from another neighborhood pooled the food they gathered each day for big group dinners. And perhaps certain communities made special deals with outlying farms to get seasonal surpluses. Fritz agrees, suggesting that families worked out distribution among themselves. As evidence, she points out that there are no remains of large storehouses that could hold piles of grain and other food for elites. There's also the cultural legacy among tribes like the Hidatsa, descended from Mississippian groups, who use the kinship system for divvying up harvests.

As urban historian William Cronon writes, a city is the sum of its buildings and its agriculture. The diversity and size of Cahokia's farms were every bit as stunning as its monumental mounds, and arguably more democratic. Though only a few people ever stood on Monks Mound to address the people, Cahokia's farms were for everyone. They were like the city's many plazas, open to the public, and providing bounty for all.

Closing up house

Cahokian celebrations didn't happen exclusively at the big mound downtown. They mostly took place in small plazas and public buildings, in neighborhoods far from the sacrifices and speeches. Not everyone could fit into the downtown Grand Plaza, of course, but these local celebrations weren't mere overflow seats. Instead, they reflected Cahokia's diverse cultural makeup. This was a city of immigrants, who didn't all share one language or set of traditions. Especially during feast times, when lots of tourists came to town, there would have been family reunions and gatherings of people who came from similar backgrounds. They would have participated in variations on the big downtown celebration, perhaps involving local leaders who could conduct ceremonies in the languages preferred by people in that neighborhood.

One such ceremony is known colloquially among archaeologists as "closing up," and it will sound familiar because it resembles what happened to Dido's house and many others at Çatalhöyük. Throughout the city, researchers have found evidence that Cahokians had special rituals for ending the life of a house or building. First they would pull up the wooden poles that formed its walls, recycling them for use as firewood. Then, they would carefully fill the postholes with colorful clay, sometimes flecked with shimmering mica, or mixed with pottery and tools that may have come from the home's past history. The floor would be washed with water or covered in soil. Sometimes a house would be burned. It's not uncommon to find that people dug a pit in the floor of an abandoned house, then filled it with the smoking remains of domestic items: pottery, corn cobs, woven mats, jewelry, broken stone knives. All these rituals seemed to seal up the old house, and made the space ready for a new home.

Cahokians loved to build their homes right on top of the ritual-istically closed-up floors of previous ones. They drove new wall posts right into the filled postholes of the last house. When archaeologists excavate a home, they frequently find several floors carefully packed atop each other, each one representing roughly a generation of residents. It's as if people were giving their houses funeral rites. You might say Cahokians believed their city was alive, but they also accepted that its lifespan was finite.

The closing up ritual isn't quite the same as what we saw at Çatal-höyük, where it appears that homes were often abandoned and left to collapse before new people moved in atop them. And it's not what we see at Pompeii, where a new generation of *liberti* converted the villas of their predecessors into bakeries and shops. Instead, it appears to have been a way of implanting the idea of abandonment into the infrastructure of the city itself. The mounds and borrow pits were made to last, but human housing was temporary.

Perhaps this idea made it easier for people to leave the city and move on. As we think through the dramatic expansion and abandonment of Cahokia, we need to keep in mind the fundamental idea of "closing up." This was not, after all, a European or Southeast Asian city. It was an American indigenous city, and its people didn't view urbanism the way their counterparts across the oceans did. They weren't necessarily aiming to create a civilization that would spread across the earth, and last forever. Maybe they thought of the city as a version of their houses, with an end built into its beginning.

CHAPTER 11

A Great Revival

To find out more about the Cahokian world, I joined an archaeological excavation there, and visited researchers in the field for two summers a row. These excavations were led by two archaeologists who specialize in Cahokian history: Eastern Connecticut State University professor Sarah Baires and University of Toledo professor Melissa Baltus. They had assistance from researcher Elizabeth Watts and many tireless undergraduates from the Institute for Field Research. Together, they spent their summers opening up three large trenches in what they thought would be a sleepy little residential neighborhood southwest of Monks Mound.

The more they dug, the more obvious it became that this was no ordinary place. The structures they excavated were full of ritual objects that had been charred by sacred fires. We found the remains of feasts along with a rare earthen structure lined with yellow soils. Baires, Baltus, and their team had accidentally stumbled on an archaeological treasure trove that was linked to the city's demise. The story of this place would take us back to the final decades of a great city where public life was undergoing a radical transformation.

East St. Louis palimpsest

Uncovering a lost city in the modern world isn't exactly like playing *Tomb Raider*. Instead of hacking through jungle and fighting a dragon, I drove to Cahokia on a road that winds through the working-class neighborhoods of East St. Louis and into Collinsville, Illinois. As recently as the 1970s, the ancient city's elevated walkways and mounds were covered over by suburban developments. Just west of Monks Mound was the Mounds Drive-In Theater. For centuries, farmers plowed over Cahokia's smaller landmarks. Mounds were demolished for construction projects. Construction workers in the 19th century demolished an enormous pyramid called Big Mound that once towered over St. Louis, using its clay as fill beneath the railroad tracks.

All that changed 40 years ago when Illinois declared Cahokia a state historic site, and UNESCO granted it World Heritage status. The state bought 2,200 acres of land from residents, clearing away the drive-in and a small subdivision. Now the Cahokia Mounds State Historic Site is devoted to preserving what remains of the ancient city's monumental downtown architecture.

By the time I got to Cahokia, Baires and Baltus had already been digging for several weeks in the broiling southern Illinois heat. To reach the excavation, I pulled up on a gravel turnout behind some old gas tanks and trudged through the muddy grass of an unmarked field until I saw a bunch of people with shovels clustered around three open pits. It was 7 a.m., but I was actually a bit tardy—the team started

every day around 6:30 a.m. to avoid working through the late afternoon heat.

Baires and Baltus chose to explore this unassuming area, known as
the Spring Lake Tract,[1] based on a magnetometry survey that Watts
had done several months before. Using a handy shoulder-mounted
magnetometer, Watts carefully paced out the entire field, looking for
signs of ancient habitation.

Magnetometers are perfect for sniffing out buried structures
because they can detect anomalies that represent disturbed earth,
burned objects, and metals several feet beneath the surface. Watts'
magnetometry map revealed a distinctive pattern of promising dark
rectangular spots or anomalies, their shapes and positions too pre-
cise to be natural. They looked an awful lot like the floors of homes
arranged in a semicircle, perhaps around a courtyard.

The courtyard shape is what caught Baltus and Baires' attention.
Late in Cahokia's history, there was an inexplicable shift in the city's
layout: people abruptly stopped building on a north-south grid and
returned to open courtyard plans that imitated the village layouts
from before Cahokia's founding. Maybe they were seeing one of those
late-era neighborhoods in the magnetometry. There was another
alluring element to the Spring Lake Tract as well. The archaeologists wanted to know what ordinary people were doing during the
city's transition, and this spot was well beyond the elite sphere of
Monks Mound.

So they broke into the earth above three separate anomalies,
eventually creating three trenches called excavation blocks (EB1,
EB2, and EB3).

As I ambled into the dig site, Baires, Baltus, and Watts were looking down at EB1, muttering to each other about what they'd found.
"Ugh—what is this?" Baires asked, looking frustratedly at the floor of

a structure that had not seen light for almost a thousand years. I knelt down next to her at the carefully squared-off edge of the pit, trying to imagine a building here. "It's a palimpsest," Watts suggested. The group had uncovered layer upon layer of material, suggesting many structures were built in this same place over time. Like most of the team, Watts stood barefoot in the muddy trench so as not to disturb the ground where Cahokians once walked.

Even with my untrained eye, I could tell she was pointing at overlapping building floors: one area of darker clay ended abruptly in a diagonal line like a wall, and alongside it was a uniformly colored area of clay studded with charcoal and artifacts. The walls themselves, made from wooden posts sunk into the clay, would have been removed and recycled by the Cahokians long ago.

EB1 was the size of a modest home, but its life had been far from ordinary. At least one ritual fire burned here, its flames consuming valuable offerings like mica, a beautifully woven mat, a pottery trowel imported from a remote village, and an ancient projectile point from pre-Cahokia peoples that would have been centuries old by the time it was buried here. EB2 and EB3 were similarly unusual, yielding finds that suggested feasting and ritualistic earth-moving activities.

What Baires and Baltus thought would be a bunch of private homes turned out to be a public area full of "special-use structures," the preferred archaeological term for any building whose purpose goes beyond the everyday. People constructed these buildings for everything from political debates and social gatherings to spiritual practices and party venues. Looking over the neighborhood, Baires said simply, "I've never seen anything like this." Following her gaze, I could no longer see the empty field bordered by trees and distant gas tanks. Instead, there were meeting halls, a wide courtyard with a decorated wooden pole at its center, and a sacred pit where Cahokians borrowed clay for their mounds. A huge trash pile full of deer bones and broken pottery hinted at a big feast.

I was looking back in time to a period when the quiet fields around me would have been packed with people, houses, and mounds all the way to the horizon.

Over the Spring Lake Tract the sky was a scalding blue, and the heat was clotted with humidity. Baires and Watts revealed their secret to staying cool: bring a bottle of completely frozen water in the morning and it will have melted to chilled perfection by midday. It's excellent for pressing against sweaty foreheads as it defrosts, too. Even though the excavation areas were shaded with canvas roofs, we took frequent breaks to guzzle water and reapply sunblock. Everyone wore hats with varying degrees of sartorial cunning. Ultimately it didn't matter how dorky you looked, as long as you didn't go home with a burned neck or face.

At first, I wandered between the excavation blocks, trailing after Baires and Baltus as they made their rounds and checked the students' work. At EB1 and EB2, there were dozens of finds: chunks of ceremonial pottery, a tiny human face re-created in clay, projectile points, the remains of a woven mat, and the triangular handle of a special beaker that once held the hallucinogenic Black Drink. EB3 remained a mystery. It looked like part of a palisade wall ringing the neighborhood on the magnetometry survey, but Baires and Baltus came to believe it might be something else.

The two crouched together at the edge of each block, conferring with Watts and the students. Occasionally they directed the students to wrap an especially valuable find in tinfoil, or fold it into a lunch bag. Everything was carefully labeled; even the soil was scooped into buckets and pushed through a sieve later to catch any remaining items.

I started to learn the verbal shorthand that Baires and Baltus had developed over years of working trenches together. Strategies for "chasing out" or "following out" features that emerged from the clay were developed on the fly. "Let's follow out this line of burned clay," Baires directed a student in EB1. The bottom of each structure was a "basin" because the Cahokians built with sunken floors. When we found a structure wall, we "caught its edge" or "caught a corner." It was as if we were racing after a history on the verge of escape.

Pretty much every dig in Cahokia begins with "chunking out" about 30 centimeters of sterile ground created by years of farmers plowing up the land. Beneath that, the city's layers begin. With each centimeter removed, the archaeologists go backward in time, working their way through the city's late-phase dissolution, into the classic era, with its masterful pottery and art. When I arrived, some of the trenches were already about a meter deep.

Digging is a specialized craft, and the students were learning it on the job. Eastern Connecticut State undergrad Emma Wink, who was working tirelessly to chase out an odd layer of yellow soil at the mysterious EB3, told me she was so focused on her work that she forgot everything else. "I'm basically a mole person," she joked. Over at EB1, where the most artifacts were emerging, Western Washington University senior Will Nolan followed out a tantalizing layer of burn. He said he could feel the difference between layers. The burn felt "crunchy, grainy, and harder to dig." He knew when he'd gone through the burn because the next layer was "smooth and sticky."

Baires loaned me a shovel with a carefully sharpened edge, and explained that I wouldn't be digging. I'd be "shovel scraping," skimming off just a thin layer of the basin at EB2. Each scrape left a curled sheet of clay in my shovel like a thick, dirty scroll. Any time I felt resistance in the mud or heard a crunch, I immediately stopped and examined the ground, using a pointed trowel to dig gently around anomalous lumps. My first find was a slab of red pottery that crumbled to dust in my fingers. "Don't worry about that," Baires assured me. "It's just unfired clay and it won't hold up." Later, I found nuggets of charcoal, blobs of yellow pigment, a few jagged pieces of fired pottery, and several burned deer bones.

The bones were the worst because there were so many of them that they halted our digging dozens of times. We had to be careful to determine that these weren't human bones because human remains must be reported immediately. Though we'd already identified ours as deer bones, the archaeologists would sometimes do a "lick check" to be sure they weren't just bone-shaped rocks. Lick check? I stared at Baires in bewilderment. "Do you want to lick it?" she asked. "Bones are porous, so your tongue will stick to it." The students looked at me. Would the weird journalist do it? Hell yeah, I would. I brought a small fragment of bone to my mouth, tasted salt, and felt my tongue adhere lightly to the surface. "Yep, that's what bone feels like," Baires said with a shrug.

After I'd been shovel scraping for an hour, blisters started coming up and popping on my fingers. Later, having fallen exhausted into bed at 8:30 p.m., I could feel the exact part of my thigh that I used to push the shovel handle. I couldn't stop thinking about how I licked the bones of a deer that had been cooked for a feast in Cahokia 900 years ago. I wished I had been there to see the party, but this might have been the next best thing.

The democratization of Cahokia

If you look online or in books for illustrations that re-create Cahokia, you'll notice an almost universal error. The mounds and swales of the city are shown covered in a light dusting of green grass, almost like a golf course. Nothing could be further from the truth. In a ground-breaking book called *Envisioning Cahokia*, a group of archaeologists explains that the city and its monuments would have been built from bald, black mud. No grass would have survived within city limits, though many houses were surrounded by gardens for growing beans, squash, and other staples.

Against the dark, swampy mud, the wood-framed and thatched houses of Cahokians were colorful, decorated with mats, carvings, and plaster. Cahokians planted wooden poles in public areas, possibly painted and decorated with fur, feathers, baskets of grain, and other symbolic items. We can't be sure whether these poles were ceremonial, or more like signposts. Perhaps they were both. Archaeologists can see where they were placed by charting the locations of deep, cylindrical holes punched into the earth of mounds, plazas, and the front yards of houses. Though the wood has long ago rotted away, its shape remains, sometimes with a bit of ceremonial mica or ochre at the bottom, tucked beneath the bottom of the post.

Archaeologists mark the eras of the city based on the orientation of its houses. During the Lohmann phase (1050–1100 CE), when people first built Cahokia's Grand Plaza and Monks Mound, they organized houses into courtyard patterns with several dwellings facing a small central plaza. During the Stirling phase (1100–1200 CE), often called Classic Cahokia, people built on a strict grid with houses and mounds oriented in a north-south direction. This was also the city's heyday, when it had the biggest population. In the final, Moorehead phase (1200–1350 CE), people returned to the courtyard plans of the Lohmann phase.

But these different city phases weren't just architectural fads. Archaeologist Susan Alt argues the transformation "marked social change." Nowhere are these changes more obvious than in the downtown area where Monks Mound rises above the Grand Plaza. This central meeting place was an engineering marvel, carefully sloped during the city's construction to allow water to drain off during public events. Everything about the architecture here suggests a highly stratified society led by charismatic figures who lived above Cahokia's sprawl on the smoothed top of Monks Mound. Ordinary residents of the city spent many long hours ritualistically hauling clay in baskets from borrow pits to build the mounds. The leaders repaid them with words of wisdom and massive feasts. But at some point that wasn't enough anymore.

During the late Stirling phase, there must have been quite a bit of urban unrest. The elites of Monks Mound erected an enormous wooden palisade wall all the way around the Grand Plaza, effectively enclosing themselves in a walled neighborhood and turning the public space of the plaza into something more private or exclusive. This may have led to more problems. If people were literally kept out of the downtown area by a giant wall, as Baltus puts it, "they might feel disenfranchised." Shortly thereafter, the Grand Plaza fell into disrepair. Alt writes, "Domestic buildings and refuse-filled features seem to have been relocated around and onto the plaza perimeter, perhaps as part of a general redesign of downtown Cahokia in conjunction with the recently built palisade wall. By 1300, there were probably few to no residents left in this inner sanctum." In other words, nonelites moved into the area and even dumped trash there. During this period, people also tore down Woodhenge, the great circle of wooden poles marking the solstice.

As the city reshaped itself during the Moorehead phase, Cahokians violently rejected the people and symbols of their once-monumental downtown. Roughly half the city's population moved away, and those who remained began to retreat into their own neighborhoods, conducting smaller public rituals and events. The courtyard and public buildings in the Spring Lake Tract reflected this new kind of social organization. Local communities had supplanted the city's central authority.

One might argue that the city went from an authoritarian design to something more democratic. Lane Fargher, an archaeologist studying the urban development of indigenous cities located in today's Oaxaca, Mexico, describes a city called Tlaxcallan that was built in the 1250s CE, during the years when Cahokia was undergoing its big revival and transition. Writing about Fargher's work in *Science*, journalist Lizzie Wade explains:

> Most Mesoamerican cities were centered on a monumental core of pyramids and plazas. In Tlaxcallan, the plazas were scattered throughout every neighborhood, with no clear center or hierarchy. Rather than ruling from the heart of the city, as kings did, Fargher believes Tlaxcallan's senate likely met in a grand building he found standing alone 1 kilometer outside the city limits. This distributed layout is . . . a sign of shared political power, he says.[2]

The layout of Tlaxcallan sounds a lot like Cahokia in the Moorehead phase, when people had turned away from the Grand Plaza and built their own smaller plazas within neighborhood courtyard communities. A plurality of plazas might point toward a democratizing current in Cahokia's public culture, too.

Against "collapse"

Like all the cities we've seen in this book, Cahokia was not static, and its remains tell a story of a culture that dynamically moved through several phases over the centuries. That's why many archaeologists today are questioning the idea that civilizations have "classical" or "peak" phases that we can contrast with a "collapse" phase. The idea of collapse comes out of the same 19th- and early 20th-century colonial traditions that brought us the idea of lost cities miraculously "discovered" by European archaeologists. Thinkers in this tradition hold that all societies progress along the same path that European civilizations took, growing bigger, more hierarchical, and more industrialized over time. Societies that don't embrace market economies are dubbed "undeveloped," and cities that stop expanding are characterized as failures whose culture has collapsed. But this perspective doesn't fit the evidence.

By the 1970s, archaeologists and urban historians had accumulated loads of evidence that urban civilizations have no set developmental pattern. Plenty of cities, including Angkor and Cahokia, are organized around nonmarket principles. Metropolitan areas expand and contract with waves of immigration over time. When a city's population breaks apart into smaller villages, that isn't a failure. It's simply a transformation, often based on sound survival strategies. The culture of that city lives on in the traditions of people whose ancestors lived there, many of whom will go on to build new cities in its image. Civilizations might cycle through a number of high-density urban phases and dispersal phases over the course of centuries.

The collapse hypothesis was nearly dead when Jared Diamond published his popular book *Collapse* in 2005. Based mostly on anecdotal evidence from cultures like the Maya and Polynesians on Easter Island, he argues that societies "collapse," or fail, when they engage in environmentally unsound practices. His argument played into a lot of myths about how cities work, including the idea that cultures are wiped out when their high-density settlements disappear.

As we've seen with the cities in this book, urban abandonment does not mean some kind of cultural death. Usually it means that city people have migrated elsewhere, bringing the values, art, and technologies of the city with them to new homes. Diamond is right to highlight environment as a contributing factor in urban dissolution, but that's only one part of the story. Abandonment is most importantly a political process.

Immediately after *Collapse* came out, many archaeologists and anthropologists scrambled to correct misconceptions and errors in Diamond's account. Anthropologists Patricia McAnany and Norman Yoffee published a volume called *Questioning Collapse*, an anthology of scholars who present hard data showing that Diamond's idea of "collapse" was scientifically unsound. They argue that civilizations like the one on Easter Island were decimated by the political process of colonialism, not poor environmental practices. And when it comes to Mayan "collapse," they point out that there are still millions of Mayans living in Mexico. Can a culture that still thrives really be said to have collapsed? Guy D. Middleton, an anthropologist who has spent his career studying social transformation, chimed in with a book called *Understanding Collapse*, in which he argued that there is never a single reason for abandonment. And, moreover, societies tend to be far more resilient than their settlements.

Today, most archaeologists who study ancient cities refuse to use the term "collapse" at all, preferring instead to describe social change. Many believe that Diamond's work has misled the public about the way civilizations truly operate. Though most prefer to provide counterevidence as a corrective, others have gotten fed up. American studies scholar David Correia published an essay about Diamond's work called simply "F**k Jared Diamond."[3] Correia calls out Diamond's "environmental determinism," which leaves out the crucial political aspects of urban transformation. Meanwhile, anthropologists David Graeber and David Wingrow take issue with the way Diamond suggests that civilizations at their peak are always hierarchical, and that those hierarchies can only be dislodged by environmental catastrophe followed by a collapse. They write:

> Jared Diamond notwithstanding, there is absolutely no evidence that top-down structures of rule are the necessary consequence of large-scale organization . . . it is simply not true that ruling classes, once established, cannot be gotten rid of except by general catastrophe. To take just one well-documented example: around 200 AD, the city of Teotihuacan in the Valley of Mexico, with a population of 120,000 (one of the largest in the world at the time), appears to have undergone a profound transformation, turning its back on pyramid-temples and human sacrifice, and reconstructing itself as a vast collection of comfortable villas, all almost exactly the same size.[4]

Here they refer to democratic architecture that's similar to what we see in Moorehead-phase Cahokia, or at Tlaxcallan in what is today Mexico. Ultimately, the point Graeber and other anti-collapse scholars are making is that there is no one path to urbanism and social complexity. More importantly, urban abandonment does not lead to social collapse. People are resilient, and our cultures can survive volcanoes and floods, even if our cities don't.

These sometimes bitter debates are ultimately about how we define public spaces and the societies that use them. Every city is an experiment with using architecture to create a public sphere, and Diamond's environmental determinist perspective suggests that this sphere collapses when people mismanage their natural resources. What he gets wrong is that the public is diverse and always changing. And often, these changes can be seen clearly in city layouts. By ignoring this capacity for change, Diamond has injected a popular nihilism into stories about city-building. He suggests some civilizations are doomed to fail, while others will inevitably succeed. Perhaps a better way to look at cities is as ecosystems whose components are always transforming, and whose boundaries expand and contract naturally. Maybe all our cities are in constant cycles of centralization and dispersal; or, if we think with our galaxy brains, they are temporary stops on the long road of human public history.

CHAPTER 12

Deliberate Abandonment

When you're excavating at Cahokia, you start to appreciate what it was like to build the mounds here a millennium ago. We shoveled clay into buckets, sweated, hydrated, and repeated. Our hands were covered in garbage and dirt. We watched the sun's path overhead to mark the time, always wary of looming storm clouds. Of course, we hadn't gone completely medieval. Baltus supplemented our personal observations with a couple of weather satellite apps on her phone. Even when the sky looked cloudless, the American Bottom could brew up a storm in less than an hour.

One afternoon, everyone's mobiles lit up with dire warnings about a dangerous hailstorm. Racing against the weather, we packed up shovels and bags with military precision. Once the dark gray clouds gathered over the Mississippi, it could start pouring within minutes. We crammed ourselves into a van and took shelter at a nearby Mexican restaurant as thunder rattled the windows and winds uprooted trees in nearby East St. Louis.

Over steaming plates of enchiladas and pitchers of frozen margaritas, I pumped the archaeologists for information about what kind of social structure united tens of thousands of people in Cahokia so long ago. What could have drawn so many people to perform backbreaking labor in the broiling humidity? My thoughts went to the charismatic leaders who led Cahokia's revival movements. "Who got to be at the top of Monks Mound?" I asked. "Was it a chief or some kind of religious leader?" From the way the archaeologists looked at each other I could tell this was a trigger question. "This is a hotly debated topic," Baltus said finally with a laugh.

Even if we imagined this revival emanating from the teachings of one person, Baires cautioned that there probably wasn't a single "chieftain" leading everyone to build their houses a certain way or line their borrow pits with colorful clay. "I don't like the idea of a chieftain," she explained. "I think power was more diverse than that. It was a heterarchy."

I rolled the unfamiliar word around on my tongue. "Heterarchy—like, a monarchy except a lot of people are in power?"

The answer turned out to be yes and no. Cahokia's heterarchy might have been a lot of different groups making decisions and governing themselves. Perhaps there were craft guilds or neighborhood associations. Already, I'd seen that the Spring Lake Tract was full of ritual items. Perhaps they had their own leadership council, too? "If Cahokia was a religious movement, people might have engaged with that on their own terms," Baltus said. "Their idea of spirituality may have come from home, not the top of the mound." In other words, ordinary Cahokians may have had their own interpretations of the city's spiritual power. They followed their own local leaders and customs as well as looking to the people on Monks Mound.

Rejecting Monks Mound

In the 1960s, when scientists were still in the habit of digging up Native American ancestors without permission, an archaeologist named Melvin L. Fowler opened up a mound. There, he found the remains of several public rituals—and over 250 human bodies—that give us a glimpse of politics and spirituality in Stirling-era Cahokia.

Fowler knew that the classical Cahokia grid was mostly aligned on a north-south axis. But there was one oddly shaped mound that didn't fit. Mound 72 is one of the city's few "ridge top mounds," meaning its rectangular body was constructed with a peaked top like a roof. And though it stood precisely south of Monks Mound, it was angled 30 degrees off the east-west axis, pointing in the exact direction of the summer and winter solstice. Fowler suspected this mound might be something special.

When Fowler and his colleagues dug, they discovered that Mound 72's ridge top was actually built over three previous mounds, each one marking a significant moment in the city's history during the 10th and 11th centuries. One of those mounds contained the bodies of 52 young women, sacrificed in some way that did not leave marks on their bones. Their bodies had been stacked in two tidy layers on top of clay platforms, then ritualistically covered over with earth. Another held the bodies of men on litters, similarly arrayed. Buried beneath thousands of pounds of clay for centuries, their skeletons were pressed as flat as flowers between the pages of a book. Stable isotope analysis of their teeth, which can pinpoint where people were born, shows these people were all local to the American Bottom.

Perhaps the most famous burial in Mound 72 contains the bodies of two people, one atop the other, in what's called the "beaded burial."

The top body was placed on a river of valuable blue shell beads and may have worn a cloak fashioned to look like a falcon. The burial included hundreds of gorgeous ceremonial projectile points, as well as piles of other valuable offerings. Alongside the beaded body were the remains of several other people, including some who had no heads. The find presented a tantalizing tableau for scientists who wondered about the spiritual and political beliefs of Cahokians.

Debates over the meaning of the beaded burial have raged in the archaeological community for decades. Initially the bead-adorned skeletons were described as male, with the top one dubbed "the Birdman." Fowler and other archaeologists assumed the Birdman was a celebrated ruler or warrior, perhaps the source for contemporary Siouan stories of the superhero Red Horn. But this interpretation has been pushed aside in the wake of a groundbreaking 2016 study by Illinois State Archaeological Survey Director Tom Emerson and his colleagues, which chronicles the first comprehensive skeletal analysis of the bodies in Mound 72. They discovered that the two people at the center of the tableau are in fact a young male and a female, suggesting a ritual of fertility. This interpretation is bolstered by the remains of other male/female pairs buried with them, as well as the 52 young women who also may have represented reproductive bounty.

Now it would seem that the beaded burial wasn't marking the grave of a great warrior or Cahokian founder. Instead, Emerson argues, we're probably seeing the remains of a public performance where people representing mythical figures were sacrificed. The city's elites may have led the performance to show their political and spiritual power, much the way their European counterparts of the same era were conducting public executions and crusades. "This scene looks more like a theatrical sacrifice rather than a burial," Emerson and his colleagues write. They suggest it might have been a pageant where the city celebrated creation and renewal. Many of the offerings, like shells, are associated with the Underworld in local Native American belief systems—and the Underworld, in turn, is connected to farming and the land's fecundity.

Sacrifices like the ones in Mound 72 may have involved joyous retellings of a creation story during the height of Cahokia's power. Perhaps Cahokia's leaders incorporated sacrifice into one of the city's awe-inspiring parties, to commemorate a fruitful harvest. But over time these mass deaths may have led to resentment, especially if decisions over life and death were in the hands of a few people ruling from on high. It's possible there was a political rebellion. Evidence from the Grand Plaza's decline would seem to support this idea. After people stop using the downtown area, they also stop engaging in human sacrifice. Maybe Cahokia's citizens toppled the regime that occupied Monks Mound and created a new social model.

Without a time machine, we'll never know exactly what the Cahokians' political struggles were about. Still, we have some hints about how they saw the world. The symbols they left behind suggest they divided the universe into an Upper World of spirits and ancestors, an Underworld of Earth and animals, and a human world in between. These worlds were not entirely separated, and the liminal spaces where they intermingled were places of great power. Images that bring worlds together are common in Mississippian art. The Upper World, represented by thunder and spirits, and the Underworld, represented by water and agriculture, are intertwined. Baires and Baltus think Cahokians used water and fire in their everyday rituals to draw the Upper World and Underworld together.

We can see the transformative power of water written into the layout of Cahokia. Though the city's mounds attract the eye, the deep borrow pits were no less important to urbanites. Left open to the elements, they filled with water on a seasonal basis. The borrow pit that provided clay for Monks Mound is so enduring that it's still filled with water to this day. Many pieces of ceremonial Ramey pottery are covered in images of water and fish, while shells filled burial mounds throughout the Mississippian world.

During the Spring Lake Tract excavation, I got a chance to see how one neighborhood sculpted water into its daily activities. Baires pointed at a deep hole the students dug at EB3, uncovering about a meter of a sloping ramp paved in yellow soils. It was obvious this yellow layer wasn't natural: it wasn't found in soil from the area and followed an exact 30-degree slope downward. Baires, Baltus, and Watts speculated that it was once the entrance to a shallow borrow pit that supplied this neighborhood with mud. We could see a history of this pit in its sediment layers. At first, the locals allowed the trough to become a seasonal pond. Later, they filled it back up with carefully layered clay, almost like they were building an inverted mound. "We caught the edge of a deliberately filled borrow pit," Baires explained with a grin. It was an incredibly unusual find, which added evidence to the idea that pits were as important to Cahokians as mounds.

Fire was even more important, especially late in the city's history. Fire could join worlds because what was burned on Earth ascended to the Upper World through smoke. Everywhere archaeologists dig at Cahokia, they find charred sacrifices. In 2013, construction workers building a freeway in East St. Louis discovered the remains of a late Cahokian neighborhood built entirely for the purpose of ritual burning. Dozens of tiny houses, full of corn and other valuables, were constructed rapidly and then torched. Nobody had ever lived in those homes. It appears that the entire neighborhood was essentially burned in effigy.

At the Spring Lake Tract, all our excavation blocks were layered with periodic burns. The group at EB1 dug up enough ground for Baires and Baltus to figure out where all its overlapping structures once stood. The lowest level was a clay floor from the Stirling phase, at the height of Cahokia's power. That floor was burned at some point and covered in another layer of clay for the floor of a later structure. In the later structure, people dug a pit into the floor, carefully lined it with a mat, and then filled it with valuables like the beaker handle and ancient Woodland projectile point. People burned the pit and its contents, too, possibly to commemorate the first burn.

I watched as Baires and Baltus gingerly used their trowels to reveal charred remains of the mat that once lined the offering pit. Its furled edge wound across the clay and looked like a crisscross pattern etched in charcoal. We weren't actually looking at the mat itself, but the impression it left behind in the earth as it smoldered. "It's nuts," Baires said. "We never find things like this."

At EB2, there were no elaborately interpenetrating layers of ritual burning, but the structure itself was an unusually large rectangle that suggested a public space rather than a home. Plus, all those burned deer bones and broken Ramey pots inside were a sure sign that some kind of celebration happened here. It was easy to imagine the ceremonial structures we'd uncovered at EB1 and EB2 standing next to a ritualistically dug trench, its floor layered with pale yellow clay.

Slowly, the layout of the neighborhood was emerging around us. This was no ordinary domestic area; people who lived here were heavily engaged in the city's political and spiritual life, conducting regular rituals. But this place also represented a trend in late classical Cahokian culture. City dwellers stopped using Monks Mound and the Grand Plaza for public performances and started conducting more rituals at home, on a smaller scale. Local identity eclipsed city identity, and the rigid city grid returned to the courtyard layout of pre-Cahokian days.

This insight also shed light on the importance of the borrow pit at EB3. It was a localized version of the giant borrow pits that supplied clay for Monks Mound, offering people in the neighborhood a constant reminder of how the Underworld intrudes into our own.

Revitalization before the fall

Baires and Baltus make a good investigative team because their areas of expertise span the city's history: Baires focuses on the classic Stirling phase, while Baltus explores the later Moorehead phase. But both are fascinated by what Baltus calls a "rejuvenation period" late in the city's life. Before it was completely abandoned in 1400, Cahokia went through a final revitalization movement. This movement might have started with a person or group suggesting a new way to live, contact with new allies, or a new relationship to agriculture and the Underworld. As a result, Cahokia was rebuilt rapidly by people seemingly on fire with belief.

They reconstructed their homes using the courtyard neighborhood layouts from Cahokia's earliest days. Baltus believes they were reexamining history and seeing it in a different light. When Cahokians dug, they often found old projectile points and other items from the Woodland peoples who lived in the area before the city was built. They treasured these items, the same way people today treasure ancient objects from Cahokia. It seems that Cahokians were celebrating the same kind of "history within history" that Ian Hodder described at Çatalhöyük. Baltus and Baires found a Woodland projectile point in the ceremonial fires buried in the layers of EB1, treated with the same reverence as Ramey pottery. It's as if Cahokians were embracing retro styles or traditional values.

In the final revitalization period, people turned their obsession with the past into a new kind of social movement. "We see a return to old practices, including a decentralized religious practice," Baltus said. But this decentralization didn't stop at city boundaries. In scattered Mississippian sites across the floodplain and the uplands, Cahokian practices slowly became unbound from Cahokia proper. Farmlands like the one at the BBB Motor Site were swallowed up by forest again. Archaeologists still find ritual burnings on floors, but none of the Ramey pottery that was so characteristic of the city's symbolism. The population of the city was draining away, and as people left, they took some of Cahokia's culture with them but left other parts behind.

During the Stirling period, Cahokians built great plazas and anchored their belief systems to the land. But during the city's last revitalization, those beliefs became unmoored from the city—perhaps due to disenchantment with the old ways, or just a renewed focus on smaller communities. Eventually, the city's precincts became so distant from each other that it would be hard to call it a unified city anymore. Its public life was falling apart. After all, Baltus explains, "if you don't unite people around an identity tied to place, with practices that keep people together, there can be fragmentation."

Environmental factors also played a part in the city's fragmentation. Some archaeologists believe the city was inundated by a massive flood from the Mississippi River that was so destructive and deadly that the survivors didn't want to stay.[1] Baires and Baltus have long been skeptical of this idea[2] and devoted part of their summer to disproving it. They invited geomorphologist Michael Kolb to take soil cores around the edges of their dig site. Using a truck-mounted device, he punched out cores that went 3 meters deep, looking for a thick layer of buried river sediments suggesting a flood. He found nothing like that at all.

Cahokia did suffer through a number of droughts, however, which would have made it difficult for the city to support a large population. Given that Cahokians' beliefs were tied to the landscape, any environmental changes would have affected them culturally, too. "There's a cycle," Baltus explained. "There's a drought and that changes people's relationship to the land, then spiritual practices change, then land use changes, spiritual practices change again, and before you know it you have fragmentation and abandonment." This process sounds like a fast-motion version of what happened at Çatalhöyük, where small acts of abandonment led to larger ones until the city stood relatively empty. Eventually Cahokia also became a place where people buried their ancestors.

Cahokia grew to such an enormous size because the structure of the city itself was part of its residents' spiritual and political worldview. But over time that centralized belief system began to crumble. When the last revitalization swept the city, people returned to the old ways. They looked to home, rather than the plaza, for their sense of identity and community. Their once-unified city was divided into many peoples who left the mounds behind.

Survivance

The abandonment of the dense urban life at Cahokia was not, as Jared Diamond might say, evidence for social collapse. Instead it was a dramatic new phase in the migration of indigenous people across the landscape. Osage anthropologist Andrea Hunter has studied the next phase in Mississippian life,[3] as Cahokia's residents fanned out over the Midwest to join many Siouan tribes. The Osage oral histories tell of a great migration that began in Ohio and paused for centuries in a place where the Missouri River branches off the Mississippi, on the land where Cahokia once stood. Eventually this migration continued again, as the people who became the Osage headed west. Hunter notes that there is strong linguistic evidence tying the Osage and other Siouan tribes to the Cahokia region. Tribes scattered across the Midwest, she writes, share the same words for "corn, gourds, squash, pumpkins, beans, cultivation, plant processing, cooking preparations, and the bow." This suggests the groups had a common origin, roughly around the time of the Woodland peoples who first started cultivating these crops. Those Woodland people followed the call of a great revival, and settled at Cahokia to build an urban farming society before eventually moving on again.

Other evidence of the Cahokia-Siouan connection comes from pieces of art found at Cahokia. Many figurines and paintings depict a figure who resembles the Siouan hero Red Horn, named for his plaited hair, dyed red and sticking up from the back of his head like a horn. Many legends, still told by the Sioux, celebrate Red Horn's skills as a warrior and gamer, as well as his intense frenemy relationships with various spirits. In one story, Red Horn celebrates a triumph by turning his earlobes into human heads (thus giving him the nickname "He Who Wears Human Faces on His Ears"); in another, he returns from death after striking a clever bargain with some spirits. Red Horn was just one of many heroes celebrated in stories that the Cahokians told. It's possible Red Horn made his first appearance at Cahokia, or that he was part of an even older story told by the Woodland people who migrated to the city.

Today, the Osage are one of many tribes whose culture and ideals were shaped by the people who abandoned Cahokia. And Cahokia is still a symbol that inspires people from many tribes across the landmasses that Europeans named North America and Canada.

Mississippian culture, with its stunning mound architecture, is a reminder of the longevity and complexity of indigenous civilizations. Coushatta-Chamorro artist Santiago X has been incorporating mounds into his work for several years.[4] In one project, called New Cahokia, he built an enormous flat-top mound covered in screens that dance with images of nature, abstractions, and indigenous performances. He's also built "burial mounds" out of Chicago Blackhawks jerseys that he burns as a protest against European appropriations of tribal identity. Santiago X calls his work Indigenous Futurism, to emphasize that indigenous culture is a part of humanity's future, and not something that collapsed long in the past.

Ohkay Owingeh author Rebecca Roanhorse writes popular fantasy novels like *Trail of Lightning*, which incorporates indigenous histories and culture. One recent novel takes place partly at Cahokia, and I caught up with her as she was in the process of writing it. Speaking from her home in New Mexico, she told me that Cahokia is important to her because she wants readers to know that "there were extensive, sophisticated cities and trade routes in the Americas before European invasion." She imagines the city as very cosmopolitan, with Iron Age technology, busy streets, pens full of animals, and a rivalry with the urbanites living in Chaco Canyon to the south. Unlike many of the archaeologists I spoke with, Roanhorse says she's not particularly focused on the spirituality of the people who lived at Cahokia. "It's important to me to say that we had governments, we had hierarchy, we had trade and technology," she mused. "These are the things that [Europeans] denied we had, and our supposed lack of them was used to justify genocide and taking our land."

In the late 20th century, Anishinaabe writer and scholar Gerald Vizenor coined the term "survivance" to describe American indigenous cultures today. Though the term is intended to be ambiguous, he sums up part of its meaning in his book *Manifest Manners*: "Survivance is an active sense of presence, the continuance of native stories, not a mere reaction, or a survivable name. Native survivance stories are renunciations of dominance, tragedy, and victimry." Like Santiago X, Vizenor looks to an indigenous future full of living cultures that are always transforming. We may not know exactly what Cahokia meant to the people who lived there, but their traditions thrive in revitalized communities, reconfigured in the wake of the political disaster that was European colonialism. As Roanhorse and other indigenous artists have pointed out, tribal cultures today survived an apocalypse and are building something new. Cahokia is part of a history of indigenous American social movements that recently took the form of protests to stop an oil pipeline on land that belongs to the Standing Rock Sioux tribe. The ancient city's political spirit survives in these kinds of movements, which are focused on how people should shape the earth.

Put another way, Cahokian public life left an indelible mark on the land. Other tribes inhabited the city's empty courtyards, and European colonists built farms and suburbs over them, but the monuments of Mississippian civilization still endure. Cahokia's story feels more vital than ever in contemporary America. People didn't migrate to the mound city just to find material wealth. They sought new kinds of spiritual and political ideas in its plazas. But not everyone in Cahokia agreed on how to put those ideas into practice. For the Mississippian culture to survive, its people had to accept that their city needed to change; that's when they abandoned it to seek out something else.

One evening, around dusk, I climbed Monks Mound to check out the view that the city's rulers once had. I climbed a long set of concrete stairs, pausing to cross a flat terrace halfway up. Special-use buildings once stood here, used by shamans and elites. When I reached the top of the mound, the sky was full of tall thunderheads, and the sunset was blood red between dark blotches of cloud that glowed sporadically with lightning. The tall grass around my ankles was blinking with fireflies, and the air was cool. Below me I could see the clean ground of the Great Plaza, emptied of its ancient public. Across the river were the lights of St. Louis, a city whose citizens recently rose up to protest police brutality in Ferguson during the birth of the Black Lives Matter movement. Those protesters walked over Cahokian land whose Great Mound had been torn down over a century before, but they continued the Mississippian tradition of questioning authority.

The thick air smelled like damp soil and farmland. With my feet atop an ancient megalopolis and my eyes on distant skyscrapers, it felt as if cities were almost a natural product of this place. The land around St. Louis has been urban for a very long time. I'm not a New Agey person, but there was something undeniably magical about that. Standing on the flattened summit, I balanced on a piece of earth that almost touched the chaotic heavens. It made sense that Cahokians believed the Underworld and Upper World met here, beneath the thunder and above the clay whose shape was forever altered by human history.

Warning—Social Experiment in Progress

I moved to San Francisco in 2000, the year the tech market crashed. As first-generation digital companies with absurd business plans bled out, I witnessed a city in the process of abandonment. Every day, hundreds of people were fired, and they left the city in droves. Fancy stores that catered to web designers and coders couldn't stay open. Shopping districts began to look like the grins of people who had been repeatedly punched in the face: each darkened shop was a missing tooth. That year during the holiday season, the downtown shopping district around Union Square looked like a trash pit. Normally the pretty square would be decked out with a giant tree and menorah, but an interminable underground construction project had turned the entire place into a gaping, muddy hole.

Even those of us who didn't work in tech caught the desolate feeling. We couldn't help but notice the city transforming before our eyes. Our neighbors, so prosperous the year before, were moving back to their small towns with nothing but chunky desktop computers and DVD collections in the backs of their cars. Brand-new Ikea desks and expensive office furniture sat on every street corner in SoMa, waiting to be adopted or fall apart. For the first time in years, rents in San Francisco stayed stable rather than rising steadily. I was working at a free weekly paper, the *San Francisco Bay Guardian*, and we had to start laying people off. Our livelihood was advertising, and the city's businesses were shrinking. I wondered whether I was foolish to stay. But I had entangled my identity with the hills and swales of this city; losing it would be like losing a limb. Plus, I was lucky enough to have a cheap room in a rent-controlled house in an unfashionable neighborhood. I decided to stick it out and hope the city would survive.

It did. In fact, San Francisco today is suffering through the opposite kind of crisis, as the population explodes and the city government struggles to remake our infrastructure to support it. The second generation of tech companies is raking in cash. Though the COVID-19 pandemic changes this calculus a little, wealthy techies are gentrifying the city, driving out working-class people and other longtime residents. Developers are transforming the grid in areas like Mission Bay, which was once full of factories and warehouses, but now welcomes artisanal ice cream shops and digital production studios.

It's easy to imagine future archaeologists excavating here, trying to figure out what social movement drove people to turn industrial production facilities into tabernas. Of course, those archaeologists will have to excavate in scuba suits, or with swimming robots, because climate change guarantees that many neighborhoods in San Francisco will be underwater in 500 years. And it wouldn't be the first time that settlements here succumbed to the coastal waters. Intrepid scientists extracting cores from the submerged city will discover that human habitation began back thousands of years before Europeans arrived. Ancient environmental changes drowned a number of indigenous villages built on the shores of a river that slowly expanded, growing into the bay that divides San Francisco from Oakland today.

When we look back at the dramatic urban histories of places like Çatalhöyük, Pompeii, Angkor, and Cahokia, we can see patterns of expansion and abandonment that emerge over centuries. But even in the span of one human lifetime, a phase of urban abandonment can morph into a renewal—or vice versa. Urban rejuvenation projects can be stopped dead by several meters of hot ash; elaborate new water infrastructure systems can turn into flood hazards. A pandemic can wreck the economy. That's one reason why it's hard to predict the future of a city based on its recent history. The anxiety I experienced in San Francisco during a single economic bust-and-boom cycle might seem like nothing in retrospect, especially if the next century brings war in the Pacific, or the earthquake that Californians call "the big one" finally hits. For the same reason, we can't assume that US cities like Detroit and New Orleans—victims of economic and natural disasters in the early 21st century—will ultimately be abandoned. In 200 years, both might be thriving megacities that look nothing like they do today. Their fates depend on political will, as well as the human labor power needed to rebuild.

Though the futures of individual cities may be uncertain, we can make predictions about the likelihood that people will abandon a city, based on evidence from urban history. On the Konya Plain in Neolithic Turkey, people from scattered villages came together to form Çatalhöyük and lived there for over a millennium. Then their city burst apart again, like a dandelion, the seeds of its culture finding purchase in small villages and other great settlements that left the earth transformed, threaded with bone. We see the same pattern at Pompeii, Angkor, and Cahokia. Though contractions in the cities' populations had different causes and effects, each was precipitated by the thorny problem of managing an enormous piece of human-built infrastructure in a constantly changing environment. Managing the humans themselves was an even bigger problem. Cities are concrete embodiments of human labor, and we can read the dissolution of their publics in the crumbling of walls, reservoirs, and plazas.

Today, cities on the coasts and islands are imperiled by chaotic weather that's becoming more likely due to climate change. In 2019, cities along the Mississippi River were flooded on an unprecedented scale,[1] harming communities as well as farms. Meanwhile, heat waves are increasing across the globe;[2] in cities they are magnified by the urban heat island effect, where temperatures rise several degrees higher than in greener areas. Sweltering temperatures also mean water infrastructures will be stressed, like Angkor's were. Wildfires will claim more cities, reducing them to ash as swiftly as Vesuvius wrecked Pompeii in 79. Los Angeles narrowly missed being eaten by the Woolsey Fire in 2018; cities throughout the West spent most of summer and fall 2020 shrouded in wildfire smoke; and across the globe in Australia, fire seasons are growing even more intense. And infectious disease outbreaks are becoming more common across the globe, some exploding into deadly pandemics. Arguably, many people living in cities today are dealing with climate and health crises that will make it more difficult to maintain infrastructure and homes.

That said, we have ample evidence from history that cities can survive in adverse environments. The people of Çatalhöyük outlasted a drought by changing their diets. Even after Angkor had been parched, then flooded, a large population persisted there for centuries, patching up infrastructure. Refugees from Pompeii moved to new cities where they enjoyed prosperity, living alongside their former neighbors. Cahokia went through multiple droughts while its city grid expanded and fragmented, but that wasn't enough to drive populations away for good.

But cities today are dealing with more than fires and floods. Globally, we're in a period of political instability and authoritarian nationalism. Unfortunately, evidence from history shows that this can be a death knell for cities. Though powerful leaders can mobilize labor for massive infrastructure projects, this kind of top-down system of urban development rarely remains stable for long. An abused labor force is an unhappy labor force, and that's how abandonments start—especially when politics steer city design, rather than sensible engineering. Troubled urban leadership can trigger a diaspora, which is what appears to have happened at Çatalhöyük, Angkor, and Cahokia. That said, we have a counterexample at Pompeii, where the government stepped in to offer humanitarian aid and disaster relief to the city's refugees. Though Pompeii had to be abandoned, its people did not walk away from Roman city life.

The combination of climate change and political instability we face in many modern cities suggests that we're heading for a period of global urban abandonment. As cities become more unlivable, people will die. The number of people who perish in floods, fires, and pandemics will swell beyond anything we've seen before, and scenes of broken cities littered with bodies will become commonplace. It's only a matter of time before another hurricane-ravaged city falls prey to a plague that can't be stopped because governments refused to spend money on rescue efforts.[3] Civil unrest and widening class divisions will exacerbate these problems. If our political systems can't address the twin problems of climate and poverty, there will be more food and water riots, as well as global wars over natural resources. The costs of city life will far outweigh the benefits, sparking mass migrations of people seeking new homes—and more international conflicts. Eventually, some of today's megacities will look like something out of a far-future science-fiction movie, full of half-drowned metal skeletons covered in incomprehensible advertisements for products we can no longer afford to make or buy.

But if we've learned anything from history, we know the death of a few cities doesn't mean the world will collapse into dystopia. We will survive the urban end times, just like so many people did when they abandoned Çatalhöyük, Pompeii, Angkor, and Cahokia. The question is, what will we do next?

Humans have been building cities for over 9,000 years, but it's only in the past few decades that the majority of us have lived in urban areas. With so many people flocking to our modern-day versions of Cahokia, cities seem inevitable—but they aren't.

After abandoning our future cities, some people may return to small-town life, like the people of Angkor and Çatalhöyük did. Often, farming is at the center of these kinds of communities, so we might see villagers of tomorrow eating locally, fueling their agricultural work by setting up off-the-grid power sources. There's another possibility, too. There were many people who left Cahokia and Çatalhöyük to become seminomadic. Post-urban people of the 21st and 22nd centuries might become nomads, living in their cars or other vehicles, forming caravans for safety. Earth may become a planet full of tiny human settlements, with cities being the exception rather than the rule. Depending on where you were born, this could be a relatively nice life. More likely, it would be an extremely difficult one, beset by the same hardships endured by farmers and nomads during the Neolithic—and made worse by global climate crisis and resource depletion.

There's also the possibility that we'll figure out a way to salvage our imperiled cities. Perhaps, like the people of Pompeii, we'll muster relief efforts that help people rebuild in new places. We might attempt to design a radically different kind of metropolis, like Domuztepe, that continues the traditions of the previous ones while incorporating new ideas. Maybe this process will lead to more sustainable cities built in places that can resist the worst effects of climate change. That might sound like a Utopian impossibility, but not if we learn from our urban failures. Looking back on Çatalhöyük, Pompeii, Angkor, and Cahokia, it's not hard to figure out what keeps a city vital: resilient infrastructure like good reservoirs and roads, accessible public plazas, domestic spaces for everyone, social mobility, and leaders who treat the city's workers with dignity. This is not such a tall order, especially when you consider that thousands of years ago, our ancestors managed to maintain healthy cities for centuries at a time.

Perhaps the most valuable lesson we can learn from the history of urban abandonment is that human communities are remarkably resilient. Cities may die, but our cultures and traditions survive. Urbanites have rebuilt after countless disasters, and put their neighborhoods back together in places far from where they started. Even after extended periods of urban diaspora, humans have returned to city-building again. Though nearly every generation believes it's living through the end times, there has never been a great civilizational collapse from which we didn't return. Instead, there has been only the long road of transformation, each generation handing off its unfinished projects to the next.

Cities are ongoing social experiments, and the remains of ancient homes and monuments are like half-erased lab notes left by our ancestors. They describe how people tried to bring diverse groups together with a shared purpose, to nourish and entertain each other, to overcome political conflict and climate catastrophe. They also describe our failures: the authoritarian slave-driving leadership, the bad civil engineering, and the laws that limited many people's access to resources. Our forebears' eroded palaces and villas warn us about how communities can go wrong, but their streets and plazas testify to all the times we built something meaningful together.

As long as we tell our urban ancestors' stories, no city is ever lost. They live on, in our imaginations and on our public lands, as a promise that no matter how terrible things get, humans always try again. In a thousand years, we'll still be working on the urban experiment. Sure, we'll fail again—but we'll also learn how to make things right.

ACKNOWLEDGMENTS

This project took years to research and complete, and along the way I've made friends, had incredible conversations with strangers, and traveled to places all over the world. I'm thankful for all of it. I'm most grateful to the researchers who took the time to talk to me about their ideas, welcoming me into their excavations and workplaces. Their names are in the pages of this book; I hope I have done justice to the breadth of their knowledge and good humor. Needless to say, all errors are my own.

Thanks also to my quick-witted editor at Norton, Matt Weiland, and to editorial assistant extraordinare Zarina Patwa. My superpowered agent Laurie Fox made it all possible. Jason Thompson created the gorgeous maps you see throughout the book—thanks, Jason!

And then there are all my long-suffering writing buddies and sundry victims who read pieces of this book and gave me valuable feedback: Charlie Jane Anders, Benjamin Rosenbaum, Mary Anne Mohanraj, David Moles, Anthony Ha, and Jackie Monkiewicz. Extra special thanks to my editors at *Ars Technica*, Ken Fisher, Eric Bangeman and John Timmer, who encouraged me to write the articles that eventually became the backbone of this book. For generalized inspiration and good role modeling, thanks to Carl Zimmer, Charles Mann, Rose Eveleth, Amy Harmon, Seth Mnookin, Deb Blum, Veronique Greenwood, Alondra Nelson, Maia Szalavitz, Maryn McKenna, Maggie Koerth, Jennifer Ouellette, and Thomas Levenson.

Most of all, thanks to Chris Palmer, Jesse Burns, and Charlie Jane Anders for going on long, hot, dirty trips with me, for tolerating my endless nerdsplaining about urban life, and for these past two decades of domestic history. I love you so much.

NOTES

Introduction: How Do You Lose a City?

1. Brendan M. Buckley et al., "Climate as a Contributing Factor in the Demise of Angkor, Cambodia," *Proceedings of the National Academy of Sciences* 107, no. 15 (April 2010): 6748–52.
2. "68% of the World Population Projected to Live in Urban Areas by 2050, Says UN," Department of Economic and Social Affairs, United Nations, last modified May 16, 2018, https://www.un.org/development/desa/en/news/population/2018-revision-of-world-urbanization-prospects.html.

Chapter 1: The Shock of Settled Life

1. Ian Hodder, ed., *The Archaeology of Contextual Meanings* (Cambridge: Cambridge University Press, 1987).
2. C. Tornero et al., "Seasonal Reproductive Patterns of Early Domestic Sheep at Tell Halula (PPNB, Middle Euphrates Valley): Evidence from Sequential Oxygen Isotope Analyses of Tooth Enamel," *Journal of Archaeological Science: Reports* 6 (2016): 810–18.
3. A. Nigel Goring-Morris and Anna Belfer-Cohen, "Neolithization Processes in the Levant: The Outer Envelope," *Current Anthropology* 52, no. S4 (2011): S195–S208.

4. D. E. Blasi et al., "Human Sound Systems Are Shaped by Post-Neolithic Changes in Bite Configuration," *Science* 363, no. 6432 (March 15, 2019).

5. Carolyn Nakamura and Lynn Meskell, "Articulate Bodies: Forms and Figures at Çatalhöyük," *Journal of Archaeological Method and Theory* 16 (2009): 205–30.

6. Ian Hodder, *The Leopard's Tale: Revealing the Mysteries of Çatalhöyük* (New York: Thames and Hudson, 2006).

7. Peter Wilson, *The Domestication of the Human Species* (New Haven, CT: Yale University Press, 1991).

8. Wilson, *Domestication of the Human Species*, 98.

9. Julia Gresky, Juliane Haelm, and Lee Clare, "Modified Human Crania from Göbekli Tepe Provide Evidence for a New Form of Neolithic Skull Cult," *Science Advances* 3, no. 6 (June 28, 2017): e1700564.

10. K. Schmidt, "Göbekli Tepe—the Stone Age Sanctuaries. New Results of Ongoing Excavations with a Special Focus on Sculptures and High Reliefs," *Documenta Praehistorica* 37 (2010): 239–56.

11. Marion Benz and Joachim Bauer, "Symbols of Power—Symbols of Crisis? A Psycho-Social Approach to Early Neolithic Symbol Systems," *Neo-Lithics Special Issue* (2013): 11–24.

12. Janet Carston and Stephen Hugh-Jones, *About the House: Lévi-Strauss and Beyond* (Cambridge: Cambridge University Press, 1995).

13. Çigdem Atakuman, "Deciphering Later Neolithic Stamp Seal Imagery of Northern Mesopotamia," *Documenta Praehistorica* 40 (2013): 247–64.

14. Hodder, *Leopard's Tale*, 63.

Chapter 2: The Truth about Goddesses

1. Kamilla Pawłowska, "The Smells of Neolithic Çatalhöyük, Turkey: Time and Space of Human Activity," *Journal of Anthropological Archaeology* 36 (2014): 1–11.

2. Ian Hodder and Arkadiusz Marciniak, eds., *Assembling Çatalhöyük* (Leeds: Maney, 2015).

3. Ruth Tringham, "Dido and the Basket: Fragments toward a Non-Linear History," in *Object Stories: Artifacts and Archaeologists*, ed. A. Clarke, U. Frederick, and S. Brown (Walnut Creek, CA: Left Coast Press, 2015).

4. Michael Marshall, "Family Ties Doubted in Stone-Age Farmers," *New Sci-*

entist (July 1, 2011), https://www.newscientist.com/article/dn20646-family-ties-doubted-in-stone-age-farmers/.

5. Nerissa Russell, "Mammals from the BACH Area," chap. 8 in *Last House on the Hill: BACH Area Reports from Çatalhöyük, Turkey*, ed. Ruth Tringham and Mirjana Stevanović, Monumenta Archaeologica, vol. 27 (Los Angeles: Cotsen Institute of Archaeology Press, 2012).

6. Michael Balter, *The Goddess and the Bull: Çatalhöyük, an Archaeological Journey to the Dawn of Civilization* (New York: Free Press, 2010).

7. Balter, *Goddess and the Bull*, 39.

8. Carolyn Nakamura, "Figurines of the BACH Area," chap. 17 in *Last House on the Hill: BACH Area Reports from Çatalhöyük, Turkey*, ed. Ruth Tringham and Mirjana Stevanović, Monumenta Archaeologica, vol. 27 (Los Angeles: Cotsen Institute of Archaeology Press, 2012).

9. Lynn M. Meskell et al., "Figured Lifeworlds and Depositional Practices at Çatalhöyük," *Cambridge Archaeological Journal* 18 (2008): 139–61; see also Carolyn Nakamura and Lynn Meskell, "Articulate Bodies: Forms and Figures at Çatalhöyük," *Journal of Archaeological Method and Theory* 16 (2009): 205.

10. Meskell et al., "Figured Lifeworlds and Depositional Practices at Çatalhöyük," 144.

11. Ian Hodder, *The Leopard's Tale: Revealing the Mysteries of Çatalhöyük* (New York: Thames and Hudson, 2006).

12. Rosemary Joyce, *Ancient Bodies, Ancient Lives: Sex, Gender, and Archaeology* (London: Thames and Hudson, 2008), 10.

13. Wendy Matthews, "Household Life Histories and Boundaries: Microstratigraphy and Micromorphology of Architectural Surfaces in Building 3 (BACH)," chap. 7 in *Last House on the Hill: BACH Area Reports from Çatalhöyük, Turkey*, ed. Ruth Tringham and Mirjana Stevanović, Monumenta Archaeologica, vol. 27 (Los Angeles: Cotsen Institute of Archaeology Press, 2012).

14. Burcum Hanzade Arkun, "Neolithic Plasters of the Near East: Catal Hoyuk Building 5, a Case Study" (master's thesis, University of Pennsylvania, 2003).

15. Daphne E. Gallagher and Roderick J. McIntosh, "Agriculture and Urbanism," chap. 7 in *The Cambridge World History*, ed. Graeme Barker and Candice Goucher (Cambridge: Cambridge University Press, 2015), 186–209.

16. Hodder, *Leopard's Tale*, chap. 6.

17. Jeremy Nobel, "Finding Connection through 'Chosen Family,'" *Psychology Today*, last modified June 14, 2019, https://www.psychologytoday.com/us/blog/being-unlonely/201906/finding-connection-through-chosen-family.

Chapter 3: History within History

1. Sophie Moore, "Burials and Identities at Historic Period Çatalhöyük," *Heritage Turkey* 4 (2014): 29.
2. Patricia McAnany and Norman Yoffee, *Questioning Collapse: Human Resilience, Ecological Vulnerability, and the Aftermath of Empire* (Cambridge: Cambridge University Press, 2009).
3. Melody Warnick, "Why You're Miserable after a Move," *Psychology Today* (July 13, 2016), https://www.psychologytoday.com/us/blog/is-where-you-belong/201607/why-youre-miserable-after-move.
4. "Immigration," American Psychological Association, accessed November 12, 2019, https://www.apa.org/topics/immigration/index.
5. Pascal Flohr et al., "Evidence of Resilience to Past Climate Change in Southwest Asia: Early Farming Communities and the 9.2 and 8.2 Ka Events," *Quaternary Science Reviews* 136 (2016): 23–39.
6. Peter Schwartz and Doug Randall, "An Abrupt Climate Change Scenario and Its Implications for United States National Security" (October 2003), accessed November 11, 2019, https://web.archive.org/web/20090320054750/http://www.climate.org/PDF/clim_change_scenario.pdf.
7. Daniel Glick, "The Big Thaw," *National Geographic* (September 2004).
8. Ofer Bar-Yosef, "Facing Climatic Hazards: Paleolithic Foragers and Neolithic Farmers," *Quaternary International* pt. B, 428 (2017): 64–72.
9. Flohr et al., "Evidence of Resilience to Past Climate Change in Southwest Asia."
10. Michael Price, "Animal Fat on Ancient Pottery Reveals a Nearly Catastrophic Period of Human Prehistory," *Science* (August 13, 2018), https://www.sciencemag.org/news/2018/08/animal-fat-ancient-pottery-shards-reveals-nearly-catastrophic-period-human-prehistory.
11. David Orton et al., "A Tale of Two Tells: Dating the Çatalhöyük West Mound," *Antiquity* 92, no. 363 (2018): 620–39.
12. Ian Kuijt, "People and Space in Early Agricultural Villages: Exploring Daily Lives, Community Size, and Architecture in the Late Pre-Pottery Neolithic," *Journal of Anthropological Archaeology* 19, no. 1 (2000): 75–102.
13. Monica Smith, *Cities: The First 6,000 Years* (New York: Viking, 2019), 9.
14. Joseph Tainter, *The Collapse of Complex Societies* (Cambridge: Cambridge University Press, 1988).
15. William Cronon, *Nature's Metropolis: Chicago and the Great West* (New York: W. W. Norton, 1991).

16. Stuart Campbell, "The Dead and the Living in Late Neolithic Mesopotamia," in *Sepolti tra i vivi. Evidenza ed interpretazione di contesti funerari in abitato. Atti del Convegno Internazionale* [Buried among the Living], ed. Gilda Bartoloni and M. Gilda Benedettini (Università degli Studi di Roma "La Sapienza," April 26–29, 2006), https://www.academia.edu/3390086/ The_Dead_and_the_Living_in_Late_Neolithic_Mesopotamia.

Chapter 4: Riot on the Via dell'Abbondanza

1. Marco Merola, "Pompeii before the Romans," *Archaeology Magazine* (January/ February 2016).

2. Mary Beard, *Pompeii: The Life of a Roman Town* (London: Profile Books, 2008).

3. "Samnite Culture in Pompeii Survived Roman Conquest," *Italy Magazine*, last modified July 6, 2005, https://www.italymagazine.com/italy/campania/ samnite-culture-pompeii-survived-roman-conquest.

4. Andrew Wallace-Hadrill, *Houses and Society in Pompeii and Herculaneum* (Princeton, NJ: Princeton University Press, 1994).

5. Translation appears in Alison E. Cooley and M. G. L. Cooley, *Pompeii and Herculaneum: A Sourcebook* (New York: Routledge, 2013).

6. Eve D'Ambria, *Roman Women* (Cambridge: Cambridge University Press, 2007).

7. D'Ambria, *Roman Women*.

8. Pliny the Elder, Book 7, Letter 24, accessed November 12, 2019, http://www .vroma.org/~hwalker/Pliny/Pliny07-24-E.html.

9. "Via Consolare Project," San Francisco State University, accessed November 11, 2019, http://www.sfsu.edu/~pompeii/.

10. Henrik Mouritsen, *The Freedman in the Roman World* (Cambridge: Cambridge University Press, 2011).

11. Mouritsen, *The Freedman in the Roman World*, 121, 140.

12. Heather Pringle, "How Ancient Rome's 1% Hijacked the Beach," *Hakai Magazine* (April 5, 2016), https://www.hakaimagazine.com/features/how-ancient-romes-1-hijacked-beach/.

Chapter 5: What We Do in Public

1. Ilaria Battiloro and Marcello Mogetta, "New Investigations at the Sanctuary of Venus in Pompeii: Interim Report on the 2017 Season of the Venus Pompei-

ana Project," accessed November 1, 2019, http://www.fastionline.org/docs/FOLDER-it-2018-425.pdf.

2. Steven Ellis, *The Roman Retail Revolution: The Socio-Economic World of the Taberna* (Oxford: Oxford University Press, 2018).

3. Miko Flohr, "Reconsidering the Atrium House: Domestic Fullonicae at Pompeii," in *Pompeii: Art, Industry and Infrastructure*, ed. Eric Poehler, Miko Flohr, and Kevin Cole (Barnsley, UK: Oxbow Books, 2011).

4. Lei Dong, Carlo Ratti, and Siqi Zheng, "Predicting Neighborhoods' Socioeconomic Attributes Using Restaurant Data," *Proceedings of the National Academy of Sciences* 116, no. 31 (July 2019): 15,447–52.

5. Eric Poehler, *The Traffic Systems of Pompeii* (Oxford: Oxford University Press, 2017).

6. Mouritsen, *The Freedman in the Roman World*, 122.

7. Classicist Beth Severy-Hoven suggests there were further signs of the brothers' uneasy relationship with their class position in some of the paintings inside their villa as well. Beth Severy-Hoven, "Master Narratives and the Wall Painting of the House of the Vettii, Pompeii," *Gender & History* 24 (2012): 540–80.

8. Sarah Levin-Richardson, "Fututa Sum Hic: Female Subjectivity and Agency in Pompeian Sexual Graffiti," *Classical Journal* 108, no. 3 (2013): 319–45.

9. Sarah Levin-Richardson, *The Brothel of Pompeii: Sex, Class, and Gender at the Margins of Roman Society* (Cambridge: Cambridge University Press, 2019).

10. Levin-Richardson, "Fututa Sum Hic."

11. Ann Olga Koloski-Ostrow, *The Archaeology of Sanitation in Roman Italy: Toilets, Sewers, and Water Systems* (Chapel Hill: University of North Carolina Press, 2015).

Chapter 6: After the Mountain Burned

1. Recent evidence suggests the eruption was in fall, rather than in late summer as had been previously thought. "Pompeii: Vesuvius Eruption May Have Been Later than Thought," BBC World News, last modified October 16, 2018, https://www.bbc.com/news/world-europe-45874858.

2. William Melmouth, trans., *Letters of Pliny*, Project Gutenberg, last updated May 13, 2016, https://www.gutenberg.org/files/2811/2811-h/2811-h.htm#link2H_4_0065.

3. Brandon Thomas Luke, "Roman Pompeii, Geography of Death and Escape: The Deaths of Vesuvius" (master's thesis, Kent State, 2013).

4. Nancy K. Bristow, "'It's as Bad as Anything Can Be': Patients, Identity, and the Influenza Pandemic," supplement 3, *Public Health Reports* 125 (2010): 134–44.

5. J. Andrew Dufton, "The Architectural and Social Dynamics of Gentrification in Roman North Africa," *American Journal of Archaeology* 123, no. 2 (2019): 263–90.

6. Andrew Zissos, ed., *A Companion to the Flavian Age of Imperial Rome* (Malden, MA: Wiley & Sons, 2016).

Chapter 7: An Alternate History of Agriculture

1. "Ancient Aliens," History Channel (May 4, 2012), https://www.history.com/shows/ancient-aliens/season-4/episode-10.

2. Patrick Roberts, *Tropical Forests in Prehistory, History, and Modernity* (Oxford: Oxford University Press, 2019).

3. Patrick Roberts et al., "The Deep Human Prehistory of Global Tropical Forests and Its Relevance for Modern Conservation," *Nature Plants* 3, no. 8 (2007).

4. Spiro Kostof, *The City Shaped: Urban Patterns and Meanings through History* (London: Thames and Hudson, 1999).

Chapter 8: Empire of Water

1. Miriam T. Stark, "From Funan to Angkor: Collapse and Regeneration in Ancient Cambodia," chap. 10 in *After Collapse: The Regeneration of Complex Societies*, ed. Glenn M. Schwartz and John J. Nichols (Tucson: University of Arizona Press, 2006), 144–67.

2. Eileen Lustig, Damian Evans, and Ngaire Richards, "Words across Space and Time: An Analysis of Lexical Items in Khmer Inscriptions, Sixth–Fourteenth Centuries CE," *Journal of Southeast Asian Studies* 38, no. 1 (2007): 1–26.

3. Zhou Daguan, *A Record of Cambodia: A Land and Its People*, trans. Peter Harris (Chiang Mai, Thailand: Silkworm Books, 2007).

4. David Eltis and Stanley L. Engerman, eds., *The Cambridge World History of Slavery*, vol. 3 (Cambridge: Cambridge University Press, 2011).

5. Lustig et al., "Words across Space and Time."

6. Miriam Stark, "Universal Rule and Precarious Empire: Power and Fragility in the Angkorian State," chap. 9 in *The Evolution of Fragility: Setting the*

Terms, ed. Norman Yoffee (Cambridge: McDonald Institute for Archaeological Research, 2019).

7. Matthew Desmond, "In Order to Understand the Brutality of American Capitalism, You Have to Start on the Plantation," *New York Times Magazine*, August 14, 2019, https://www.nytimes.com/interactive/2019/08/14/magazine/slavery-capitalism.html.

8. Stark, "Universal Rule and Precarious Empire."

9. Stark, "Universal Rule and Precarious Empire."

10. Kenneth R. Hall, "Khmer Commercial Development and Foreign Contacts under Sūryavarman I," *Journal of the Economic and Social History of the Orient* 18, no. 3 (1975): 318–36.

11. Dan Penny et al., "Hydrological History of the West Baray, Angkor, Revealed through Palynological Analysis of Sediments from the West Mebon," in *Bulletin de l'École française d'Extrême-Orient* 92 (2005): 497–521.

12. Christophe Pottier, "Under the Western Baray Waters," chap. 28 in *Uncovering Southeast Asia's Past*, ed. Elisabeth A. Bacus, Ian Glover, and Vincent Piggot (Singapore: National University of Singapore Press, 2006), 298–309.

13. Penny et al., "Hydrological History of the West Baray, Angkor."

14. Monica Smith, *Cities: The First 6,000 Years* (New York: Viking, 2019).

15. Saskia Sassen, "Global Cities as Today's Frontiers," Leuphana Digital School, https://www.youtube.com/watch?v=Iu-p31RkCXI. She also elaborates on these ideas in her book *The Global Cities: New York, London, Tokyo* (Princeton, NJ: Princeton University Press, 1991).

16. Geoffrey West, *Scale: The Universal Laws of Life, Growth, and Death in Organisms, Cities, and Companies* (New York: Penguin, 2018).

17. Lustig et al., "Words across Space and Time"; see also Eileen Lustig and Terry Lustig, "New Insights into 'les interminables listes nominatives des esclaves' from Numerical Analyses of the Personnel in Angkorian Inscriptions," *Aséanie* 31 (2013): 55–83.

18. Kunthea Chhom, *Inscriptions of Koh Ker 1* (Budapest: Hungarian Southeast Asian Research Institute, 2011), https://www.academia.edu/14872809/Inscriptions_of_Koh_Ker_n_I.

19. Terry Leslie Lustig and Eileen Joan Lustig, "Following the Non-Money Trail: Reconciling Some Angkorian Temple Accounts," *Journal of Indo-Pacific Archaeology* 39 (August 2015): 26–37.

20. "Household Archaeology at Angkor Wat," *Khmer Times*, July 7, 2016, https://www.khmertimeskh.com/25557/household-archaeology-at-angkor-wat/.

21. Lustig and Lustig, "Following the Non-Money Trail."
22. Eileen Lustig, "Money Doesn't Make the World Go Round: Angkor's Non-Monetization," in *Economic Development, Integration, and Morality in Asia and the Americas*, ed. D. Wood, Research in Economic Anthropology, vol. 29 (2009), 165–99.
23. Lustig, "Money Doesn't Make the World Go Round."
24. Mitch Hendrickson et al., "Industries of Angkor Project: Preliminary Investigation of Iron Production at Boeng Kroam, Preah Khan of Kompong Svay," *Journal of Indo-Pacific Archaeology* 42 (2018): 32–42, https://journals.lib.washington.edu/index.php/JIPA/article/view/15257/12812.
25. Damian Evans and Roland Fletcher, "The Landscape of Angkor Wat Redefined," *Antiquity* 89, no. 348 (2015): 1402–19.

Chapter 9: The Remains of Imperialism

1. Henri Mouhot, *Travels in the Central Parts of Indo-China (Siam), Cambodia, and Laos during the Years 1858, 1859, and 1860*, 2 vols., Gutenberg Project, last modified August 11, 2014, http://www.gutenberg.org/files/46559/46559-h/46559-h.htm.
2. Alison Carter, "Stop Saying the French Discovered Angkor," *Alison in Cambodia* (blog), accessed November 12, 2019, https://alisonincambodia.wordpress.com/2014/10/05/stop-saying-the-french-discovered-angkor/.
3. Terry Lustig et al., "Evidence for the Breakdown of an Angkorian Hydraulic System, and Its Historical Implications for Understanding the Khmer Empire," *Journal of Archaeological Science: Reports* 17 (2018): 195–211.
4. Keo Duong, "Jayavarman IV: King Usurper?" (master's thesis, Chulalongkorn University, 2012).
5. Tegan Hall, Dan Penny, and Rebecca Hamilton, "Re-Evaluating the Occupation History of Koh Ker, Cambodia, during the Angkor Period: A Palaeo-Ecological Approach," *PLoS ONE* 13, no. 10 (2018): e0203962, https://doi.org/10.1371/journal.pone.0203962.
6. Kunthea Chhom, *Inscriptions of Koh Ker 1* (Budapest: Hungarian Southeast Asian Research Institute, 2011), https://www.academia.edu/14872809/Inscriptions_of_Koh_Ker_n_I, 12.
7. Eileen Lustig and Terry Lustig, "New Insights into 'les interminables listes nominatives des esclaves' from Numerical Analyses of the Personnel in Angkorian Inscriptions," *Aséanie* 31 (2013): 55–83.

8. Lustig et al., "Evidence for the Breakdown of an Angkorian Hydraulic System."

9. Wensheng Lan et al., "Microbial Community Analysis of Fresh and Old Microbial Biofilms on Bayon Temple Sandstone of Angkor Thom, Cambodia," *Microbial Ecology* 60, no. 1 (2010): 105–15, doi:10.1007/s00248-010-9707-5.

10. Peter D. Sharrock, "Garuḍa, Vajrapāṇi and Religious Change in Jayavarman VII's Angkor," *Journal of Southeast Asian Studies* 40, no. 1 (2009): 111–51.

11. Roland Fletcher et al., "The Development of the Water Management System of Angkor: A Provisional Model," *Bulletin of the Indo-Pacific Prehistory Association* 28 (2008): 57–66.

12. Dan Penny et al., "The Demise of Angkor: Systemic Vulnerability of Urban Infrastructure to Climatic Variations," *Science Advances* 4, no. 10 (October 17, 2018): eaau4029.

13. Solomon M. Hsiang and Amir S. Jina, "Geography, Depreciation, and Growth," *American Economic Review* 105, no. 5 (2015): 252–56.

14. Alison K. Carter et al., "Temple Occupation and the Tempo of Collapse at Angkor Wat, Cambodia," *Proceedings of the National Academy of Sciences* 116, no. 25 (June 2019): 12226–31.

15. Dan Penny et al., "Geoarchaeological Evidence from Angkor, Cambodia, Reveals a Gradual Decline Rather than a Catastrophic 15th-Century Collapse," *Proceedings of the National Academy of Sciences* 116, no. 11 (March 2019): 4871–76.

16. Miriam Stark, "Universal Rule and Precarious Empire: Power and Fragility in the Angkorian State," chap. 9 in *The Evolution of Fragility: Setting the Terms*, ed. Norman Yoffee (Cambridge: McDonald Institute for Archaeological Research, 2019), 174.

Chapter 10: America's Ancient Pyramids

1. Sarah E. Baires, *Land of Water, City of the Dead: Religion and Cahokia's Emergence* (Tuscaloosa: University of Alabama Press, 2017).

2. See Michael Hittman, *Wovoka and the Ghost Dance* (Lincoln: University of Nebraska Press, 1997), and Alice Beck Kehoe, *The Ghost Dance: Ethnohistory and Revitalization* (New York: Holt, Rinehart and Winston, 1989).

3. John Noble Wilford, "Ancient Indian Site Challenges Ideas on Early American Life," *New York Times*, September 19, 1997, https://www.nytimes.com/1997/09/19/us/ancient-indian-site-challenges-ideas-on-early-american-life.html.

4. Timothy Pauketat, *Cahokia: Ancient America's Great City on the Mississippi* (New York: Viking, 2009).

5. Rinita A. Dalan et al., *Envisioning Cahokia: A Landscape Perspective* (DeKalb: Northern Illinois University Press, 2003).

6. V. Gordon Childe, "The Urban Revolution," *Town Planning Review* 21, no. 1 (1950): 3–17.

7. Dalan et al., *Envisioning Cahokia*, 129.

8. Timothy Pauketat, "America's First Pastime," *Archaeology* 6, no. 5 (September/October 2009), https://archive.archaeology.org/0909/abstracts/pastime.html.

9. The painter George Catlin wrote in a letter that he'd watched the Siouan Mandan tribe playing the game in the 1830s. From George Catlin, *Letters and Notes on the Manners, Customs, and Conditions of North American Indians*, no. 19, retrieved November 12, 2019, https://user.xmission.com/~drudy/mtman/html/catlin/letter19.html.

10. Margaret Gaca and Emma Wink, "Archaeoacoustics: Relative Soundscapes between Monks Mound and the Grand Plaza" (poster presented at the 60th Annual Midwest Archaeological Conference, Iowa City, Iowa, October 4–6, 2016).

11. Thomas E. Emerson et al., "Paradigms Lost: Reconfiguring Cahokia's Mound 72 Beaded Burial," *American Antiquity* 81, no. 3 (2016): 405–25.

12. Baires, *Land of Water, City of the Dead*, 92–93.

13. Andrew M. Munro, "Timothy R. Pauketat, *An Archaeology of the Cosmos: Rethinking Agency and Religion in Ancient America*," *Journal of Skyscape Archaeology* 4, no. 2 (2019): 252–56.

14. Gayle Fritz, *Feeding Cahokia: Early Agriculture in the North American Heartland* (Tuscaloosa: University of Alabama Press, 2019), 89.

15. Fritz, *Feeding Cahokia*, 150.

16. Natalie G. Mueller et al., "Growing the Lost Crops of Eastern North America's Original Agricultural System," *Nature Plants* 3 (2017).

17. Fritz, *Feeding Cahokia*, 146.

18. Fritz, *Feeding Cahokia*, 143.

Chapter 11: A Great Revival

1. Sarah E. Baires, Melissa R. Baltus, and Elizabeth Watts Malouchos, "Exploring New Cahokian Neighborhoods: Structure Density Estimates from the Spring Lake Tract, Cahokia," *American Antiquity* 82, no. 4 (2017): 742–60.

2. Lizzie Wade, "It Wasn't Just Greece—Archaeologists and Early Democracy in the Americas," *Science* (March 15, 2017), https://www.sciencemag.org/news/2017/03/it-wasnt-just-greece-archaeologists-find-early-democratic-societies-americas.

3. David Correia, "F**k Jared Diamond," *Capitalism Nature Socialism* 24, no. 4 (2013): 1–6.

4. David Graeber and David Wingrow, "How to Change the Course of Human History," *Eurozine* (March 2, 2018), https://www.eurozine.com/change-course-human-history/.

Chapter 12: Deliberate Abandonment

1. Samuel E. Munoz et al., "Cahokia's Emergence and Decline Coincided with Shifts of Flood Frequency on the Mississippi River," *Proceedings of the National Academy of Sciences* 112, no. 20 (May 2015): 6319–24.

2. Sarah E. Baires, Melissa R. Baltus, and Meghan E. Buchanan, "Correlation Does Not Equal Causation: Questioning the Great Cahokia Flood," *Proceedings of the National Academy of Sciences* 112, no. 29 (July 2015): E3753.

3. Andrea Hunter, "Ancestral Osage Geography," in Andrea A. Hunter, James Munkres, and Barker Fariss, *Osage Nation NAGPRA Claim for Human Remains Removed from the Clarksville Mound Group (23PI6), Pike County, Missouri* (Pawhuska, OK: Osage Nation Historic Preservation Office, 2013), 1–60, https://www.osagenation-nsn.gov/who-we-are/historic-preservation/osage-cultural-history.

4. Margaret Carrigan, "One Mound at a Time: Native American Artist Santiago X on Rebuilding Indigenous Cities," *Art Newspaper*, September 29, 2019, https://www.theartnewspaper.com/amp/interview/native-american-artist-santiago-x-on-rebuilding-indigenous-cities-one-mound-at-a-time.

Epilogue: Warning—Social Experiment in Progress

1. Sarah Almukhtar et al., "The Great Flood of 2019," *New York Times*, September 11, 2019, https://www.nytimes.com/interactive/2019/09/11/us/midwest-flooding.html.

2. Kendra Pierre-Lewis, "Heatwaves in the Age of Climate Change," *New*

York Times, July 18, 2019, https://www.nytimes.com/2019/07/18/climate/heatwave-climate-change.html.

3. Annalee Newitz, *Scatter, Adapt, and Remember: How Humans Will Survive a Mass Extinction* (New York: Doubleday, 2013).

INDEX

Note: Page numbers in *italics* indicate maps.